love and science

a memoir

love and science

a memoir

JAN VILCEK

Seven Stories Press

NEW YORK • OAKLAND

A SEVEN STORIES PRESS FIRST EDITION

Seven Stories Press
140 Watts Street
New York, NY 10013
sevenstories.com

Library of Congress Cataloging-in-Publication Data

Vilcek, J., 1933- , author.
  Love and science : a memoir / Jan Vilcek. -- First edition.
      p. ; cm.
  ISBN 978-1-60980-668-2 (hardback)
  I. Title.
  [DNLM: 1. Vilcek, J., 1933- 2. Physicians--Slovakia--Personal Narratives. 3. Physicians--United States--Personal Narratives. 4. Microbiology--Slovakia--Personal Narratives. 5. Microbiology--United States--Personal Narratives. WZ 100]
  R154.B623
  610.92--dc23
  [B]
                                        2015025051

Printed in the United States of America

9 8 7 6 5 4 3 2 1

"How great the merit, and the bliss how sweet,
When in fond union love and science meet."

Independent Gazetteer (Philadelphia), April 24, 1790

Contents

PART ONE | # The Wonder of Science

Chapter One

The Cure for Cancer that Wasn't

Scientists love abbreviations and acronyms. What better way to peacock one's expertise than by mystifying others with your area's secret codes? When I go to a lecture given by a scientist working in a field different from my own I often get lost among the acronyms. And we immunologists are as afflicted with this tendency as anyone.

IFN is the abbreviation for interferon—a family of natural proteins produced in an organism, usually in response to an infection. First identified in the late 1950s by the London-based British virologist Alick Isaacs and his Swiss colleague Jean Lindenmann, interferons play important roles in the defense against viruses and other infectious agents, and in the regulation of immune functions.

TNF—which stands for tumor necrosis factor—was identified by Lloyd Old and his colleagues at the Memorial Sloan Kettering Cancer Center in New York City in the mid-1970s as a protein produced in experimental animals injected with bacteria or bacterial components. The name derives from the observation that the factor appeared to cause the death of tumor tissues, or, put more scientifically, to produce "tumor necrosis."

The work that led to the identification of TNF was an outgrowth of older studies showing that bacterial infections in humans or in laboratory animals would sometimes lead to a shrinking and, in very rare cases, even complete disappearance of malignant tumors. A similar shrinking of tumors was seen in tumor-bearing experimental animals injected with low doses of toxins derived from some bacteria. However, these earlier observations left unanswered the question of whether the shrinking of tumors was a direct result of the action of the bacterial toxins or whether it was perhaps mediated by something made in the body in response to the toxins. Lloyd Old's study suggested that TNF, a protein produced mainly by white blood cells, was the mediator responsible for the regression of tumors. The implication was that TNF was part of the body's defense system against tumors. The study also raised the prospect that—when isolated and properly defined—the TNF protein might one day become useful as a therapeutic agent in the fight against cancer.

I first met Lloyd—then a rising star in the emerging tumor immunology field—shortly after I had joined the NYU School of Medicine as an assistant professor in the Department of Microbiology in 1965. I was intrigued by TNF from the outset because it was a natural protein produced in the body that like interferon—a protein I had worked on since the late 1950s—appeared to have a role in the immune system's array of natural defenses.

By the mid-1970s it was becoming apparent that there existed a large number of secreted proteins that were important in the fine-tuning of the body's immune responses. In 1974 immunologists agreed that secreted proteins whose primary function is to regulate immune responses, such as interferon and TNF, be called cytokines. The first half of the term, originating from the Greek *kýtos*, meaning "cell," was inspired by the fact that these proteins are both derived from cells and act on cells. The latter half of the word—"kine," as in "kinetic"—implies that the function of these proteins is to move the immune system into action.

A quick computer search of the published biomedical literature on PubMed—a comprehensive database of the US National Library of Medicine—reveals more than six hundred thousand printed scientific publications where the word "cytokine" has been used. I commiserate with the medical and science students who have to learn about the hundreds of cytokines that have been discovered in recent decades. Cytokines tend to be identified by acronyms (like TNF, IFN, and many others) or by the abbreviation IL—for interleukin—followed by a serial number, starting with IL-1. At this point we are up to IL-38, but the actual number is much larger because many interleukins consist of several molecular variants.

Some years after the original publication by Lloyd Old, my interest in TNF became more tangible. In the early 1980s, we were in my NYU laboratory using cells isolated from human blood to generate a type of interferon called IFN-gamma. Soon we realized that some other unknown cytokines were produced together with IFN-gamma in the same test tubes.

The methods available for the identification of cytokines in those days were still cumbersome and relied on the use of indirect biological assays. With my colleagues Donna Stone-Wolff, Hanna Kelker, and others, we eventually established that the fluids harvested from cultures of white blood cells, which served as the source of IFN-gamma, also contained two other cytokines: one of these we identified as a protein known among immunologists as lymphotoxin, and the other—though at the time defying definitive identification—we suspected of being identical to TNF.

In the summer of 1982, while attending a meeting on cytokines held on the campus of Haverford College in Pennsylvania, I ran into Michael Wall, a biotechnology entrepreneur whom I had known since the late 1970s. I had first met Michael when he visited me at my NYU laboratory. He was a principal at a tissue culture supply company called Flow Laboratories—a company he had founded but later sold. Remaining with Flow Laboratories after its sale, Michael was looking for opportunities to expand into biotechnology. He came to see me because he heard about our work with interferon and was interested in establishing a collaboration with my laboratory.

Michael, an MIT-trained electrical engineer, impressed me with his grasp of the biomedical field. We also hit it off personally. He struck me not only as a man with a passion for entrepreneurship, but also as someone who cared deeply about science, in addition to being a lively and charismatic person with wide-ranging interests. We agreed to strive to establish a collaboration. I had several subsequent meetings with him and his professional colleagues, but before the collaboration could get off the ground Michael decided to leave Flow Laboratories in order to pursue other opportunities. One opportunity Michael seized was the establishment of Centocor—a company that would come to play an important role in my work and life.

The story of Centocor stands out in the history of the biotechnology industry. The company was built around one technology—monoclonal antibody production. The undisputed original creators of this technology are Georges Köhler and César Milstein, for which they would earn the Nobel Prize in Physiology or Medicine in 1984. Invention of the monoclonal antibody technology represented the realization of the German immunologist Paul Ehrlich's dream of a "magic bullet," a compound that could selectively target a harmful agent. Before this groundbreaking development, antibodies could be generated only in live animals, and the resulting "polyclonal" antibodies found in the blood would represent a mixture of thousands of molecules with different specificities and properties.

Even though the potential commercial relevance of the original work by Köhler and Milstein done at Cambridge University was apparent to many from the beginning, the British National Research Development Corporation failed to file a patent application for the technology with the justification that it was impossible to identify immediate commercial uses.

Others were not so shy. One person who recognized the potential of the monoclonal antibody technology was Hilary Koprowski, a colorful, prominent Polish-born virologist and longtime director of the Wistar Institute in Philadelphia. Koprowski, with some of his colleagues, adapted the Köhler and Milstein technology for the production of antibodies directed against viruses and tumors, filing patent applications in the process. Within a short time, in 1979, Koprowski joined forces with Michael Wall to create the company that became Centocor. The company established its headquarters near Wall's home in Malvern, Pennsylvania.

When I ran into Michael again at Haverford College in the summer of 1982, the company had just moved into its first laboratories and offices.

Characteristically, Michael was full of enthusiasm and optimism about his new venture. Would I want to come to visit his new place and meet some of his collaborators? I did. At the time Centocor was still a very small enterprise—I estimate that they had some fifty employees, including a handful of scientists. I remember meeting Hubert Schoemaker, a Dutch-born molecular-biologist-turned-entrepreneur with a PhD from MIT who had become Centocor's president and CEO. Michael was the company's chairman.

My first visit to Centocor was largely ceremonial, but I did speak to Michael and Hubert about their aspirations for the company. They envisioned the venture to be fully based on applications offered by the monoclonal antibody technology, with the aim of developing diagnostic products, which took less time to bring to market than therapeutics.

To keep their costs down, Hubert and Michael said, the company would not be depending on in-house research for the development of its products. Instead, Centocor would license new technologies from universities or other partners. In implementing this strategy, Centocor, along with the rest of the biotechnology industry, was helped by the passing of the Bayh-Dole Act in 1980, which allowed universities and research institutions to patent and commercialize US government–funded research without having to pay royalties to the government. The US biotechnology industry owes much of its worldwide success to this one legislative act.

"By the way," Michael asked, "is there something your laboratory is doing that would be of interest to us?" Indeed there was. I told him about our work on IFN-gamma. With the help of Junming "Jimmy" Le, who was about to join my lab, we were planning to generate monoclonal antibodies to IFN-gamma. Although we were thinking of producing the antibodies mainly as a tool for our own research into the nature of IFN-gamma, the antibodies could also be used for the detection of IFN-gamma and its quantitative measurement. Having such a test (referred to as an "assay") for the identification and quantification of IFN-gamma in biological samples would have potential diagnostic applications; for example, the presence of IFN-gamma in some body fluids or tissues might indicate immunity or sensitization to a component of a microbial agent. Michael and Hubert seemed to like the idea and we agreed to stay in touch.

My visit to Malvern was followed by more detailed discussions about a joint project between my lab at NYU and Centocor. Soon the discussions progressed to an actual collaboration. As I had proposed, the collaboration initially centered on the development of an assay for IFN-gamma. The basis for the assay were two monoclonal antibodies specific for IFN-gamma, generated by Jimmy Le in my laboratory. By the early 1980s, the technology of monoclonal antibody production was well established and quite widely employed. Jimmy, then a recent arrival from Shanghai, had mastered the technology prior to joining my laboratory.

Once we established that the IFN-gamma assay was working in principle, we proceeded to discuss the terms of an agreement between NYU School of Medicine and Centocor. In those days, collaborative agreements between academic institutions and pharmaceutical or biotechnology companies were still relatively rare. Today NYU Medical Center has a large technology transfer office that routinely handles these types of negotiations. In 1983, there was no specialized office and the negotiations were conducted by the associate dean, Dr. David Scotch, a physician-turned-manager who bore most of the administrative burden of the entire medical center.

To provide legal advice during the negotiations, NYU engaged Peter Ludwig, a patent attorney based at a private law firm. (As happens to be the case with many of my professional contacts, Peter has become a lifelong friend.) Hubert Schoemaker—business-savvy, but eminently fair and delightful to deal with—represented Centocor in the negotiations.

NYU and I wished to accomplish two goals. First, we wanted to secure financial support for our research that would allow completion of the work we were planning to do jointly with Centocor and also provide my laboratory with some extra funds that we could ferret away for new, more adventurous projects. Second, we wanted to make sure that if a commercially successful product emerged from the project, NYU would be paid appropriate royalties on the sales of the product. (I should mention that NYU—like other universities in the US—had and continues to have rules in place for sharing a portion of the royalty payments with faculty members or other employees who contribute to the invention.)

The first goal was relatively easy to accomplish. Centocor agreed to provide funding for research in my laboratory, based on a simple research agreement signed by the two parties. A more complicated and detailed license agreement did not get signed until mid-1984, though we had already begun our scientific collaboration under the terms of the research agreement. One of the issues was the scope of the project. The initial plan was to cover only the use of monoclonal antibodies to IFN-gamma, but I was arguing for a broader agreement.

The agreement was being negotiated at a time when my colleagues and I had come to realize that several cytokines were cogenerated with IFN-gamma in the human white blood cell cultures we used for the preparation of IFN-gamma. We had just completed the publication of a paper showing that a factor termed lymphotoxin and another cytokine, likely related to TNF, were produced in the cultures along with IFN-gamma. I argued that to understand the biological functions of IFN-gamma it was important to pay attention to its interactions with these other agents. We were eager to try to generate monoclonal antibodies to some other cytokines because we believed they would be invaluable tools for the dissection of cytokine functions.

To justify the broadening of the project, in a proposal submitted to Centocor titled "Monoclonal Antibodies to Interferons and Cytokines," I wrote:

> The availability of MoAbs [abbreviation for "monoclonal antibodies"] to Lymphotoxin and Monocyte Cytotoxin would be useful for laboratory studies and would help to determine the relationships among Lymphotoxin, Monocyte Cytotoxin and TNF. In addition, since some of these molecules might play a role in autoimmune disorders as well as in natural resistance to malig-

> nancies, such MoAbs could become useful for diagnostic and
> other medical applications.

Arguing that "such MoAbs could become useful for diagnostic and other medical applications" without offering any specific examples of how the antibodies could be utilized was pretty vague. Had I submitted a similar proposal to an established pharmaceutical company, it would very likely have been dismissed with a chuckle.

Licensing agreements, especially for potential therapeutic drugs, are usually signed when there is clear supporting evidence for the utility of a specific product, such as extensive data from studies in experimental animals. With the exception of the antibodies to IFN-gamma, we not only did not have the products, we did not even know precisely what the products were going to be or what they might be used for.

It is a testament to Michael Wall and Hubert Schoemaker's trust and risk-tolerance that they accepted my proposal and agreed to sign a licensing agreement between NYU and Centocor, stipulating that my laboratory would provide Centocor with monoclonal antibodies to several cytokines, including IFN-gamma, lymphotoxin, and TNF. In return, Centocor agreed to provide research support for my laboratory for three years (eventually Centocor ended up supporting our research for fifteen years), to exert their best effort to develop products, and to pay royalties to NYU on the sales of any products based on monoclonal antibodies originating in my laboratory.

———

In December 1984, I attended a workshop, the first of its kind, for a small group of devotees interested in TNF and related factors, organized by Lloyd Old and his colleagues at the Memorial Sloan Kettering Cancer Center in New York City. In those days, the group of people actively interested in this field was still small. In fact, the approximately twenty-five attendees from America, Europe, and Asia present at the workshop represented most of the world's scientists working on TNF at the time. Only a few years later, similar meetings would be attended by hundreds of participants.

One important reason why interest in TNF was so modest in 1984 was that the factor was still poorly characterized and difficult to produce and identify. However, the New York workshop marked the beginning of a dramatic change when, seemingly out of nowhere, Bharat "Bart" Aggarwal, then a young protein chemist working at the biotechnology company Genentech, reported the purification of the human TNF protein and unveiled its complete amino acid sequence to the workshop participants. Each of the many thousands of proteins in the body, including the TNF protein, is made up of chains of tightly linked amino acids. Twenty different types of amino acids are known and their positioning in the chain—referred to as the amino acid sequence—determines the shape and function of each protein.

Less than a year after the 1984 workshop, Genentech scientists Aggarwal, David Goeddel, and their colleagues published not only the complete amino acid sequence of the TNF protein but also the sequence of the DNA encoding the TNF protein, along with details of the organization and chromosomal location of its gene. This information—in addition to having broad scientific interest—formed the basis for the production of human TNF protein by recombinant DNA technology, thus for the first time making pure TNF protein available for scientific studies.

To explain, recombinant DNA is created by combining the genetic sequence encoding a protein—such as the TNF protein—with some other DNA sequences that, upon insertion into a living cell (bacterium, yeast, or animal cell), will direct the synthesis of the desired protein. The protein can then be isolated from the producing cells or their environment, purified, and "bottled." Much of the biotechnology industry has been built on advances in the production of proteins by recombinant DNA technology.

Progress in the molecular characterization of the TNF protein and of the corresponding DNA sequence transformed a small, insular field of investigation into an exciting discipline that almost overnight became accessible to rigorous scientific inquiry. Genentech's scientists were generous in providing investigators at academic institutions with free samples of pure recombinant TNF protein (meaning the TNF protein produced by recombinant DNA technology) for experimental studies. Of course, before giving the protein away, patent attorneys representing Genentech would make sure that the rights for commercial applications of TNF were properly secured.

In 1985, when we received the first gift of recombinant human TNF from Genentech, virtually nothing was known about the spectrum of TNF's biological actions and its molecular underpinnings. The two biological activities then known to be associated with TNF were those that had originally been identified by Lloyd Old's group: the ability to shrink tumors in animals and to kill some tumor cells grown outside the body in test tubes.

There was one additional function that came to light immediately after the publication of TNF's protein sequence. Anthony "Tony" Cerami's lab at the Rockefeller University in New York City (located across the street from Lloyd Old's lab) had independently been studying a factor dubbed "cachectin," suspected of causing the wasting in animals infected with the parasitic agent *Trypanosoma brucei*, the cause of African sleeping sickness. In a fascinating turn of events, when the cachectin protein had been isolated and sequenced, it became clear that it was identical to TNF.

Thus two laboratories, located a few hundred yards from one another, and for years focusing their work on separate projects, had come to realize that the same protein was the cause of the two seemingly unrelated phenomena they studied. Cerami's evidence that cachectin/TNF can act as a mediator of disease symptoms presaged many subsequent demonstrations of TNF's role in the genesis of disease.

———

Having received a supply of recombinant TNF from Genentech—the very first time we could lay our hands on pure TNF protein—we needed to decide what experiments we would use it for. We felt like kids in a candy store—what should we try first?

The project we had decided to embark on initially was aimed at solving the quandary of why TNF was selectively toxic for tumor cells while sparing normal cells. It was known that cytokines generally act on cells by binding to "receptors," meaning that cytokine proteins contain "keys" that fit specific "keyholes" on the surface of responsive cells. Was the selectivity of TNF action on tumor cells perhaps caused by the fact that appropriate receptors existed only on tumor cells, but not on normal ones?

To answer this question a visiting scientist from Japan, Masafumi "Masa" Tsujimoto, compared the presence of TNF receptors on the surface of tumor cells that were known to be susceptible to killing by TNF with normal human FS-4 fibroblasts (cells derived from tiny foreskins removed from newborn baby boys by circumcision that we had for many years used as a source of interferon), in which TNF was not known to produce cell damage. Using radioactively labeled TNF, Masa found that specific cell surface TNF receptors were present on both tumor cells and FS-4 fibroblasts. This finding indicated that the selective killing of tumor cells by TNF could not be explained by a difference in the presence of receptors on tumor cells and normal cells.

Now that we had identified TNF receptors on normal fibroblasts, it would be logical to expect that TNF elicited some actions in these cells. But what were they? We did not have to look for long. After exposing cultures of FS-4 fibroblasts grown in test tubes to infinitesimally small amounts of TNF, we readily observed a change in the shape of the cells under an ordinary light microscope. In the presence of TNF, cells became elongated and they also grew faster, and as a result they became more "crowded" inside the test tubes. I remember how stunned I was by this observation, because, until then, I knew of no other natural protein that would cause such a striking change in the appearance of cells.

One of my graduate students, Vito Palombella, then took on the task of showing that TNF could indeed promote the growth of normal human fibroblasts—an unexpected finding at a time when TNF was thought to be a protein that caused selective killing of tumor cells.

More surprises followed. Masayoshi "Yoshi" Kohase, another visiting scientist from Japan, and Luiz Reis, a graduate student from Brazil, were instrumental in establishing that under some circumstances TNF inhibited virus replication in a manner somewhat similar to interferon. Today we know that TNF, along with interferon, is important in the defense against virus infections.

So surprising was the multitude and breadth of TNF actions that Jedd Wolchok, an MD-PhD student in the laboratory (whose last name would sometimes get confused with mine), proposed—only half-jokingly—that the abbreviation TNF should stand for "too numerous functions." All of these findings, together with parallel findings made by colleagues in other laboratories, led to our present understanding of TNF as a cytokine with a broad range of activities affecting the immune system and other functions in the body.

Another project with surprising results was initiated by Tae Ho Lee, a graduate student from Korea. At the time it was already known that exposure of cells to TNF leads to gene activation resulting in the synthesis of a number of proteins that are not produced in the absence of TNF. Tae Ho, who received training in molecular biology while working for a Korean biotechnology company, agreed to attempt to identify yet unknown TNF-activated genes.

Today, such a project would be quite routine, as there are now established tools and automated equipment available for this type of work, and it is common to engage a service laboratory to carry out the relevant analyses. At the time, such tools did not yet exist, and Tae Ho had to employ laborious manual methods to identify, one by one, DNA sequences corresponding to cellular genes that are turned on by TNF. Tae Ho's hard work has paid off: two sequences he identified represented previously unidentified genes. For a molecular biologist to find a new gene is, I imagine, a thrill akin to identifying a new plant or animal species, or a new planet.

We decided to focus our subsequent efforts on one of the two newly identified genes, which we termed TNF-stimulated gene 6 (or TSG-6) because it was the sixth consecutive genetic sequence found to be TNF-inducible in Tae Ho's experiments. After establishing some fundamental properties of the TSG-6 gene and the protein encoded by it— referred to as TSG-6 protein—we tried hard to define the function of TSG-6 protein.

Before completing his PhD training, Tae Ho was joined in the TSG-6 project by Hans-Georg Wisniewski. Georg came to my lab as a post-doctoral fellow from East Germany just a few months before the fall of the Berlin Wall and collapse of the East German state. When Tae Ho returned to his native Korea, Georg took over the TSG-6 project. He has now spent twenty-five years analyzing the properties and functions of TSG-6 protein, a project he is still actively pursuing. Perhaps not unexpectedly, TSG-6 protein has turned out to be important in understanding innate immunity and inflammation—processes known be intrinsic to TNF actions. Unexpectedly, however, TSG-6 has also been found to be important in female fertility. We are still hoping that one day there will be practical medical applications stemming from our research on TSG-6.

———

One important reason why scientists, including our colleagues at Genentech, had so eagerly pursued the purification and characterization of TNF was the hope that the TNF protein—believed to selectively kill cancer cells while inflicting no harm on normal cells—might prove useful in the treatment of cancer. When pure recombinant TNF became available, several leading medical centers started to make preparations for the clinical evaluation of TNF's possible worth in treating cancer.

Disappointingly, the very first clinical studies revealed that—even at doses too low to produce tumor regression—humans were exceedingly susceptible to TNF's toxic effects that included a severe drop in blood pressure, blood clot formation, and adverse impacts on the heart muscle. Eventually, these findings put an end to plans for TNF's use as a therapeutic agent. Genentech's investment in the study of TNF brought the company prestige and visibility, but no marketable products. Fortunately, Genentech has since developed many successful therapeutic products to offset this—and some other—failures.

As one set of studies provided information about the toxicity of TNF given to patients by injection, other investigations showed that TNF produced within the body—for example in response to bacterial infection—played a role as a mediator of disease. The first condition in which the disease-producing role of TNF was clearly demonstrated was in the pernicious effect of a bacterial toxin in experimental animals. Bruce Beutler and Tony Cerami at Rockefeller University showed that animals injected with a lethal dose of the toxin could be protected and kept alive by injecting them with specific antibodies to TNF, thus showing that the generation of TNF in the animals' bodies was responsible for the toxin's deadly effect. Soon it became apparent that TNF was also acting as a mediator of disease in some forms of malaria and in a complication that can occur after bone marrow transplantation—graft-versus-host disease—in which the transplanted donor cells attack the recipient's organs.

In the mid-1980s, when these findings were being made and reported, the news of TNF's harmful effects came as a surprise to the scientific community. Until then, TNF—and cytokines in general—were considered to be essential for the regulation of immune responses and for boosting host defenses against infectious agents and malignant tumors. It was now becoming clear that, even though TNF plays a useful role in host defenses, when produced in excess amounts for extended periods of time, it would become harmful. As William Shakespeare noted in *As You Like It*, there can indeed be "too much of a good thing."

The realization that excess production of cytokines can be harmful, even deadly, should not have come as a complete surprise. Much earlier, Ion Gresser, an American scientist working in Paris, had conclusively shown that interferon production can have deleterious effects in animals during some virus infections.

With the indication that TNF was too toxic to be used as a therapeutic agent in humans and the findings indicating TNF caused diseases in some cases, the plans for the possible clinical exploitation of TNF had to be thoroughly rewritten. Instead of considering the administration of artificially produced TNF to patients with cancer, there was a gradual shift toward the belief that it might be more productive to consider developing agents that block the harmful disease-producing actions of TNF generated *inside* the patient's body. At this point I realized that the licensing agreement NYU had signed with Centocor could become more valuable than I had originally anticipated.

Chapter Two
....

From TNF to Remicade

Our initial collaboration with Centocor was devoted to the development of a quantitative immunological method for the detection of human IFN-gamma, a useful if not hugely exciting project. Two monoclonal antibodies produced by Jimmy Le in my laboratory proved suitable for their intended purpose. In 1985, Centocor completed the development work and the "immunoassay" was put on the market. Unlike the earlier available assays for IFN-gamma that relied on the detection of biological activity, the immunoassay was much more accurate and reproducible, and it did not require highly specialized expertise to run.

From my point of view, the assay was successful because it made it possible to detect and quantify IFN-gamma more accurately and faster than before. Centocor, however, was looking for products that could be used more widely for clinical diagnostic purposes and not just for research applications. Preliminary discussions Centocor had conducted with the federal Food and Drug Administration (FDA) about approving the assay for clinical diagnostic applications were not promising because at the time there was no clinical condition in which determining the presence of IFN-gamma would serve a useful purpose. (In addition to controlling the use of medications, the FDA also approves and monitors medical diagnostic procedures.) Besides, by the late 1980s, Centocor had begun diminishing its focus on the diagnostics arena and instead was starting to plan to develop therapeutic products that could generate higher revenues for the company. After about three years Centocor decided to discontinue the production and marketing of the IFN-gamma immunoassay.

This was also a period when Centocor was undergoing a large expansion of its workforce, mainly due to the effort to develop a monoclonal

17

antibody for the treatment of sepsis. Sepsis—"blood poisoning" in lay language, the result of severe infection—is a grave, potentially fatal condition that manifests itself mainly by decreased blood pressure, increased heart rate, and rapid breathing. The condition can progress to multiple organ failure (septic shock) and death. Sepsis and septic shock are treated with intravenous antibiotics, the administration of fluids, and general support therapy aimed at preventing or repairing organ failure; in spite of these therapies mortality is high, around 30–40 percent in patients with sepsis and up to 70 percent in patients with septic shock.

While severe infection with many microorganisms can lead to sepsis, a significant percentage is caused by types of bacteria known by microbiologists as gram-negative bacilli that contain an "endotoxin" in their membrane. In chemical terms endotoxins are lipopolysaccharides (LPS), consisting of a lipid molecule linked to a polysaccharide (complex sugar).

In 1986 Centocor initiated a major effort aimed at developing a monoclonal antibody specific for LPS into a therapeutic agent for sepsis. Expectations were that upon injection in patients suffering from sepsis caused by gram-negative bacteria, the antibody, dubbed Centoxin, would latch onto LPS, render it innocuous, and thereby halt the progression of sepsis and the development of septic shock.

Centocor management and some people in the financial community were optimistic that Centoxin would become successful. The company's eternally confident president and CEO, Hubert Schoemaker, was quoted in the press saying that by the year 2000 Centocor would become "a Merck." As much as I liked Hubert, I thought that his optimism was likely a bit of the biotech hype I'd grown accustomed to hearing.

As I already mentioned, it had been suspected since 1985 that the cytokine TNF played a role in the pathogenesis of sepsis caused by LPS, because in that year Bruce Beutler (later to become a Nobel Prize winner for unrelated studies), while working in Tony Cerami's lab at the Rockefeller University, showed that injection of a serum containing antibodies to TNF protected mice from toxic and lethal effects of bacterial LPS. To confirm the role of TNF in sepsis, in a subsequent study these scientists showed that administration of pure TNF to rats elicited disease symptoms very similar to those seen in animals with experimental sepsis.

While Centocor was focusing on the development of Centoxin, I kept urging Michael Wall and Hubert Schoemaker to keep in mind that an antibody to TNF might prove a viable supplement or alternative to Centoxin and that, in any case, an anti-TNF antibody would likely work in types of sepsis not caused by LPS-containing bacteria, which would not be responsive to Centoxin. I also pointed out that even though sepsis was not explicitly mentioned as a possible therapeutic target, the generation and commercial development of anti-TNF monoclonal antibodies was part of the existing NYU-Centocor licensing agreement.

Actually, the generation of monoclonal antibodies to human TNF ended up taking us longer than anticipated because, initially, we did not have sufficient quantities of recombinant human TNF needed for the immunization of mice. The amount of protein needed for immunizing animals—generally a prerequisite for the creation of monoclonal antibodies—is greater than the amount needed for studies in tissue culture.

It was only toward the end of 1988, when we secured the needed amounts of recombinant TNF, that Jimmy Le, in my laboratory, immunized mice and succeeded in producing mouse monoclonal antibodies to human TNF. One of the antibodies, designated "A2," not only bound TNF with high affinity and selectivity but also showed a potent neutralizing activity for TNF, a prerequisite for any antibody to be considered for treating patients.

———

Initially, Centocor seemed in no rush to commit to the development of the A2 antibody. They had their hands full with work on Centoxin, which tied up most of their resources. However, with time, Schoemaker and Wall warmed up to the idea, realizing that a therapeutic antibody to TNF would complement their existing effort to find a treatment for sepsis.

Before making a decision to proceed with the development of A2,

Centocor's management and its scientists wanted to make sure that our A2 antibody was indeed suitable for the task. By then other laboratories had succeeded in producing mouse monoclonal antibodies to human TNF. As a condition of proceeding with the project, Centocor decided that they wanted to compare the A2 antibody to the antibodies made by others. They made it clear that if some other antibody proved to have characteristics superior to A2, they would consider licensing and developing the superior antibody.

A total of four monoclonal antibodies to human TNF from three different sources were available for testing. Their potency and efficacy were compared in several laboratory tests performed independently by Centocor scientists and by us. When the testing was completed, we were pleased to learn that our A2 antibody had the most desirable characteristics. This was a fortunate outcome because we still had only one single antibody that had met the necessary criteria: to bind TNF with a high affinity and inactivate the biological function of TNF by preventing the binding of the TNF molecule to its receptor ("keyhole") on the surface of cells.

A great deal of additional work had to be done before the A2 antibody could be tested in patients. We knew that monoclonal antibodies made by mouse cells were not suitable for administration to humans, especially when repeated injections of the antibody were intended. This is because an antibody made in mouse cells is recognized as a foreign protein by the human body, leading the body to try to eliminate it. In addition to its rapid clearance from the blood stream, injection of a mouse protein into the human body can result in allergic reactions and an immune response directed at the foreign protein.

One way to circumvent the problem was to modify the original mouse monoclonal antibody by converting it into a chimeric antibody. Like chimera—the mythological creatures containing body parts from multiple animals—chimeric antibodies retain a small portion of the original mouse protein sequences, while, with the aid of genetic manipulation, the bulk of the molecule is replaced with human sequences that are not recognized as foreign by the human body. The technology, introduced in the 1980s, eliminates most of the potentially immunogenic portions of the molecule without altering its specificity for the intended therapeutic target.

Converting a mouse monoclonal antibody into a chimeric antibody was not a trivial matter at the time, but fortunately the molecular biology group at Centocor, headed by John Ghrayeb, had the necessary expertise and managed to create a chimeric counterpart of the original A2 antibody within approximately six months. A long series of preclinical evaluations in cell cultures and animal models followed, which demonstrated that the chimeric antibody, designated "cA2," had the characteristics expected of a therapeutic antibody. cA2 was the first chimeric monoclonal antibody directed at human TNF, and Centocor made sure that appropriate patent applications were filed, covering the rights to this invention as broadly as possible. Since the creation of the chimeric antibody from the NYU-generated mouse A2 antibody was accomplished by Centocor scientists, patent rights to cA2 are owned jointly by NYU and Centocor.

Subsequent testing of the cA2 antibody was conducted partly by Jimmy Le in my laboratory and partly by a group of dedicated and competent Centocor scientists including Peter Daddona, David Knight, David Shealy, and Scott Siegel. Many types of preclinical studies were needed in order to prepare an Investigational New Drug (IND) application for the FDA.

Among the experimental data usually required for the permission to conduct a clinical trial are studies showing that the experimental drug is effective in animal models. The problem with cA2 was that its action was very specifically directed against human TNF, and not TNF made in mice, rats, dogs, or other animal species. The only animals that made TNF reactive with cA2 were chimpanzees—human's closest relatives. But for ethical, financial, and logistical reasons chimps are difficult to use as experimental animals. In the end, Centocor scientists used only three chimps to demonstrate that the administration of cA2 produced no toxic side effects.

To circumvent the problem of the strict species specificity of cA2 and still satisfy the FDA requirement of showing efficacy in animals, we used mice that were genetically altered such that cells in their bodies would continually produce human TNF. Because TNF is toxic, the animals would sicken, developing a state of general physical wasting and dying by the age of about ten weeks. We showed that administration of cA2 reversed the toxic effects of human TNF in these mice and prevented their death. This was an important piece of preclinical evidence proving that cA2 was effective in the bodies of experimental animals.

These and numerous other results of preclinical studies were used for the submission of an IND application to the FDA seeking approval for the use of cA2 in clinical trials in patients with sepsis. In due course the permission was granted.

A clinical trial of cA2 in patients with sepsis was initiated by Centocor in 1991. One group of twenty patients was enrolled in the first stage of the trial aimed mainly at determining whether administration of the drug would cause toxicity. A second group of sixty sepsis patients was used to evaluate the therapeutic efficacy of cA2; those patients were given either a placebo (a dummy injection of saline) or one of three different doses of cA2. This part of the trial was "double-blind." Vials were coded so that neither the physicians nor the patients knew whether they were treated with placebo or one of the doses of cA2.

When the code was broken and the results were evaluated, it turned out that even though there were no toxic effects noted in patients receiving cA2, no clear therapeutic benefit could be demonstrated. I was not part of the team monitoring the results of the clinical trial, and I learned about them after a delay, as the bad news took some time to filter through the clinicians to Centocor and then to us. Needless to say, this was a very disappointing outcome. The failure to demonstrate efficacy in the sepsis trial seemed to deliver a fatal blow to hopes for cA2's clinical utility.

The bad news about cA2's failure to help sepsis patients was followed by more bad news for Centocor, as Centoxin—the product the company had been banking on—was also beginning to run into rough waters. Although Centoxin had been approved in Europe and recommended for approval in the US by an FDA advisory panel a year earlier, in April 1992 the FDA reversed the advisory panel's recommendation, ruling that there was insufficient evidence to support Centoxin's efficacy and that more trials would be needed before approval could be considered.

Centocor's stock, which had traded at fifty dollars a share in the beginning of 1992, dropped to nineteen dollars within hours of the news, and some months later it plunged to six dollars a share. Soon layoffs would reduce the number of Centocor employees from around 1,600 (including a highly optimistic 275-person sales force specifically hired for the distribution of Centoxin) to 400. A year later, work on Centoxin development would be completely abandoned. At that point Centocor's survival was hanging by a thread. Hubert Schoemaker stepped down as president and CEO, but retained the position of chairman. David Holveck, until then head of Centocor's diagnostic division, became the new president.

In 1994, just as the storm from the Centoxin debacle was receding, Hubert, at age forty-four, was diagnosed with medulloblastoma, a highly malignant brain tumor. After brain surgery, aggressive chemotherapy, and radiation treatment he recovered sufficiently to resume his role as chairman of Centocor. In fact, he continued to work almost as hard as before his illness, and remained the very soul of the company for the next five years. (Hubert would leave Centocor in 1999 to start a new biotech company, Neuronyx. Sadly, he succumbed to sequelae of his brain tumor and its aggressive therapy at age fifty-five. Anne Faulkner, Hubert's affectionate wife, invited me to visit him at their home in the Pennsylvania countryside about a month before his passing. During the visit he was, as ever, the perpetual optimist.)

Centocor did survive, thanks to two developments. First, the company had acquired the rights to ReoPro, a chimeric monoclonal antibody directed against a receptor on the surface of platelets, developed by Barry Coller, then working at the SUNY Stony Brook School of Medicine, and now physician in chief and vice president for medical affairs at Rockefeller University. ReoPro showed promise as a therapy aimed at preventing complications in patients who had suffered a heart attack and were undergoing angioplasty.

In 1992 Centocor sublicensed ReoPro to Eli Lilly and Company, thus securing a badly needed infusion of cash. ReoPro proved successful in clinical trials and was approved by the FDA in 1994. ReoPro was only the second monoclonal antibody approved by the FDA for clinical use. The very first monoclonal antibody approved in 1986, Orthoclone OKT3, was a mouse antibody used for short-term immunosuppression in patients who had received an organ transplant. (A complete and fascinating account of the history of monoclonal antibodies and their use in medicine can be found in Lara Marks's book *The Lock and Key of Medicine: Monoclonal Antibodies and the Transformation of Healthcare*, recently published by Yale University Press.)

The second reason for Centocor's survival was that our cA2, though not effective in sepsis, quite unexpectedly proved highly successful in other conditions. It had been suspected for some time that cytokines, including lymphotoxin and perhaps TNF, might be involved in the pathogenesis of some autoimmune disorders, which I had alluded to when drafting the proposal for our collaboration with Centocor back in 1984. Autoimmune diseases, such as rheumatoid arthritis, Crohn's disease, and many others, arise when the immune system attacks tissues and organs within its own body.

More specific insights into the role of TNF in autoimmunity emerged from studies of rheumatoid arthritis—a serious inflammatory autoimmune disorder afflicting about one out of 160 individuals in the US, and impacting women almost three times as often as men. (Rheumatoid arthritis should not be confused with osteoarthritis, a more common degenerative form of joint disease, occurring in many older people, which is not an autoimmune disorder.) Some investigators had observed that joint fluids and joint tissues of patients with rheumatoid arthritis contained numerous cytokines that were known to be associated with inflammation, including TNF.

To investigate the interaction among different cytokines and try to understand whether any of them might be a therapeutic target, Marc Feldmann, a physician and immunologist, together with his rheumatologist colleague Ravinder "Tiny" Maini at the Kennedy Institute of Rheumatology in London, took tissues from the joints of patients with rheumatoid arthritis and placed them in a test tube. Their work with this relatively simple experimental model led to the concept of a TNF-dependent cytokine cascade, in which TNF was postulated to be the prime mover responsible for the generation of many cytokines. They concluded that even though inflammation is likely caused by multiple cytokines, blocking TNF alone might be effective in arresting the inflammatory process occurring in rheumatoid arthritis because of TNF's unique role at the very top of the inflammatory pyramid. In a 1989 publication, Feldmann and colleagues proposed that in rheumatoid arthritis anti-TNF agents "may be useful in treatment."

In 1991, a team of Greek scientists based in Athens, headed by George Kollias, published a study showing that mice, genetically altered in a way that would make their bodies produce a form of human TNF, developed a severe inflammation of their joints. The joint lesions in these mice resembled the lesions in the joints of humans afflicted with rheumatoid arthritis. The authors concluded, "Our results indicate a direct involvement of TNF in the pathogenesis of arthritis." There was also later evidence from several other laboratories showing that in a model of inflammatory arthritis in mice, administration of antibodies to mouse TNF decreased the severity of joint inflammation.

Armed with these laboratory findings, Feldmann and Maini set out to find support for a small pilot clinical trial to test their idea that injecting a monoclonal antibody that blocks TNF action could be beneficial to patients with rheumatoid arthritis. They knocked on the doors of several well-established pharmaceutical and biotechnology companies but found no receptive ears.

To most scientists and physicians it seemed inconceivable that blocking a single cytokine could be beneficial for rheumatoid arthritis patients, when it was known that multiple cytokines are involved in the inflammatory process. Also, the argument went, rheumatoid arthritis is a chronic disease process, and even if short-term administration of an antibody to TNF were to prove helpful, any relief would almost certainly be temporary, as it seemed unlikely that the antibody could be administered repeatedly for long periods of time.

As you'll recall, at the time, in 1992, there was only one FDA-approved monoclonal antibody on the market—Orthoclone OKT3, used for short-term immunosuppression in patients who had received an organ transplant. Many other monoclonal antibodies had been tested for a variety of indications, mainly cancers, without apparent success. Finally, most experts then believed that blocking endogenous TNF would almost certainly compromise the immune system and lead to unacceptable side effects, including a severely decreased resistance to infections and tumors.

Feldmann and Maini could not persuade any other companies to provide them with antibodies to TNF, so they eventually turned to Centocor. Given the company's financial problems, it is doubtful that Centocor would have been willing to grant their request if it weren't for the fact that Centocor's scientific director at the time, James "Jim" Woody, was a former colleague of Feldmann's. It was this personal connection and Jim Woody's foresight that helped persuade Centocor to provide Feldmann and Maini with enough cA2 for a preliminary "open" clinical trial (meaning not a controlled double-blind trial) in twenty patients with severe rheumatoid arthritis who had failed all available treatments and had no other therapeutic options. Patients included in the study were all gravely incapacitated.

Though we have since become good friends, I had known Marc Feldmann only casually at the time and was not part of the planning of the trial at the Kennedy Institute of Rheumatology in London, carried out in May 1992 at the neighboring Charing Cross Hospital, where Maini was the head of rheumatology. When, in a conversation with Hubert Schoemaker, I heard that Feldmann and Maini's use of cA2 in rheumatoid arthritis patients produced highly promising results, I was not convinced, suspecting that Hubert's description of how dramatically the patients improved after cA2 administration might have been due to his aforementioned enthusiasm and optimism. My reservations receded when, during one of my regular visits to Centocor in Malvern, I was shown a video of a patient's response to the cA2 trial.

As I recall, the video was only about five minutes long. It showed a woman with severe rheumatoid arthritis who before the injection of cA2 could barely make it down a flight of stairs. She moved extremely slowly, carefully holding onto the handrail, and was in visible pain. Within a short time after the treatment, the woman ran down the same stairs, smiling, with ease.

The improvement seen in a majority of the other nineteen patients upon cA2 administration was also impressive. Even though this was not a controlled trial and everyone involved was well aware of the possibility of a "placebo effect"—referring to the known fact that many patients get better because they *believe* they are being helped by the treatment— the recovery of the patients was so profound and dramatic as to make it unlikely that it could be ascribed to a placebo effect.

Objective assessment of the degree of joint inflammation and joint function in the patient shown in the video corroborated the beneficial effect of cA2, as did observations and laboratory data from the remaining nineteen cA2-treated patients who showed a significant decrease in the amount of inflammation in the body. Yet I knew that experts would not be convinced unless the possibility of a placebo effect was completely ruled out. It was essential that the results be confirmed in more rigorously controlled clinical studies.

A subsequent placebo-controlled, double-blind clinical study, involving a larger group of seventy-three patients with active rheumatoid arthritis, fully validated the findings of the earlier open trial. The results, published in 1994 in the widely circulated British medical journal *The Lancet*, showed significant improvement in nearly 80 percent of cA2-treated patients, while less than 10 percent of patients treated with a placebo showed a similar improvement. This was a hugely impressive outcome. A much smaller difference between treated and untreated groups of patients would still be significant. The article concluded, "The results provide the first good evidence that specific cytokine blockade can be effective in human inflammatory disease and define a new direction for the treatment of rheumatoid arthritis."

The experience with the first groups of patients showed that even though cA2 suppressed the inflammation responsible for the symptoms of arthritis, the treatment had not cured the disease because after the initial, often dramatic, response to cA2, patients relapsed, usually within twelve to fifteen weeks after cA2 administration. Would readministration of cA2 to patients who had initially responded to treatment, but later relapsed, be effective?

As Feldmann and Maini reported in another article published in *The Lancet*, the answer to this question was yes. This was significant because cA2 would have a chance to become successful as a therapeutic agent only if it produced sustained improvement, and it was already clear that a single administration caused only transient relief. However, even with Feldmann and Maini's new data, it was still unclear how long cA2 could be administered without complications.

More good news was coming. In 1993 a group in the Netherlands headed by the gastroenterologist-scientist Sander van Deventer published a short note in *The Lancet* describing the successful treatment with a single dose of cA2 of a young girl who at age twelve was diagnosed with a severe case of Crohn's disease—a form of inflammatory bowel disease, which can affect the entire gastrointestinal tract, often accompanied by bloody diarrhea, abdominal pain, anorexia (because patients stop eating to avoid getting sick), fever, and severe weight loss.

The young patient had been treated for more than a year with all medications commonly used in Crohn's disease (prednisone, mesalazine, and azathioprine, among others), but showed no improvement. Immediately after receiving an injection of cA2, the girl's condition improved dramatically—she gained over six pounds in ten weeks and her intestinal lesions healed. The improvement lasted for three months after the single injection, after which the symptoms recurred. A follow-up open-label study with cA2 revealed that eight of ten Crohn's disease patients showed a dramatic improvement after a single injection of cA2.

With prospects high for cA2's success in the treatment of two important diseases, Centocor had to decide on which of the two conditions—rheumatoid arthritis or Crohn's disease—they should focus their efforts to obtain regulatory approval. As a more common disease, rheumatoid arthritis represented a potentially larger market. But there were other considerations that led to the decision to first seek approval for Crohn's disease.

Crohn's disease was first described in 1932 by a New York–based gastroenterologist, Dr. Burrill B. Crohn. We don't know whether the disease had not existed prior to the twentieth century or if it had been missed until then. We do know that the incidence of the disease has been increasing steadily since its original description, especially in the industrialized world.

One reason why Centocor decided to first seek approval for Crohn's disease was that there were fewer treatment options available for Crohn's disease than for rheumatoid arthritis, as no new treatments for Crohn's had been approved for three decades. The second consideration was that Crohn's disease would likely qualify for "orphan disease status," which means that fewer patients than usual could be enrolled in the clinical trials that needed to be completed for the application for FDA approval. The United States' Orphan Drug Act includes rare diseases and also diseases for which there is no reasonable expectation that the cost of developing and making available a drug will be recovered from its sales. Centocor took the position that Crohn's disease qualified for orphan disease status because, even though there were around five hundred thousand cases in the US, only about two hundred thousand of these were severe enough to require treatment with cA2. Centocor scientists thought that getting a foot in the door with the FDA would be a more straightforward process with Crohn's and that approval for Crohn's might pave the way for subsequent approvals for other indications.

Clinical trials represent a major portion of the cost of drug development, and even though the price tag was significantly less in the 1990s than it is today, trying to keep the number of enrolled patients down was an important consideration, especially for a small company like Centocor that had still not recovered from the setback suffered by Centoxin's failed regulatory approval.

The strategy turned out to be successful. Centocor sponsored two placebo-controlled "randomized trials" (another term used in the clinical trial business, meaning that patients are randomly assigned to different treatment groups) with cA2 in Crohn's disease; one included a group of 108 patients with moderate-to-severe, treatment-resistant Crohn's disease, the other focused specifically on ninety-four Crohn's disease patients suffering from enterocutaneous fistulae—a grievous complication in which perforations spanning the intestine and the abdominal wall result in the draining of the intestinal contents to the body surface. Both trials showed significant benefits from cA2 administration.

These results formed the basis for the orphan drug application submitted to the FDA in December 1997. The company requested, and was indeed granted, a fast-track regulatory review; in view of the scarcity of treatment options for Crohn's disease the application was also designated for priority review, which meant that a decision would be made within four to six months, rather than the usual ten to twelve months.

By then cA2 had been given the trade name Remicade and the generic name infliximab.

All drugs are given a trade name and generic name. The trade name is supposed to be used only when reference is made to the specific product of the company that owns the rights to it. The generic name carries over when another company produces the same product, usually after the expiration of patents. For commercial reasons drug companies tend to come up with easily pronounceable, "sexy" trade names, while generic names are often more complicated to pronounce and to remember.

In May 1998 the FDA Advisory Committee, consisting of experts in the fields of gastroenterology and immunology, held a meeting in which a decision would be made on the merits of cA2 for the treatment of Crohn's disease. The meeting was held in a large room at the Holiday Inn in Bethesda, Maryland, near the home of the FDA and the National Institutes of Health (NIH). Even though FDA advisory committee meetings are always open to the public, they are usually attended only by the advisory committee members, FDA officials, and company representatives. This time, however, the room was filled to capacity with attendees, including representatives of the media and Wall Street analysts, in addition to a handful of Crohn's disease patients.

When the morning session opened, several Crohn's disease patients who had been enrolled in the clinical trials came forward to relate the incapacitating nature of the disease. They were largely young people, the demographic most affected by the disease, which has a huge impact on their quality of life. One patient was a twenty-three-year-old woman who vividly described the debilitating nature of her fistulizing Crohn's disease and how it made living a normal life impossible, her voice trembling with emotion during her testimony. The patients also explained in impassioned language how treatment with Remicade made it possible for them to regain—at least for a short time—the ability to live a normal life. "Don't take this treatment away from us," they pleaded.

The whole room fell completely silent during the patients' testimonies, reflecting the impact they had on the audience. In subsequent reports about the meeting, Centocor representatives were quoted as saying that prior to the meeting they had no inkling that any patients would be attending; I am not so sure about that. In any case, it was certainly appropriate for the patients to attend and give independent testimony.

Following the patients' testimonials and presentations by Centocor staff and consultants that summarized the results of the clinical trials, FDA representatives raised critical comments about some aspects of the trials. Although, in general, they had not challenged the claims of significant benefits of the treatment, they disagreed with some interpretations of the data. For example, an FDA official differed with Centocor's conclusions concerning the duration of the beneficial effect of Remicade in patients with enterocutaneous fistulae. Another concern voiced by the FDA was whether Remicade was perhaps responsible for the observation of a higher incidence of lymphomas, a type of blood cancer affecting cells that are part of the immune system. Although the number of patients was too small to reach a definitive verdict, the FDA concluded that a relationship between Remicade and the development of lymphomas could not be ruled out.

After the conclusion of the presentations, advisory committee members exchanged heated arguments about the merits of the therapy. Many questions were left unanswered. The most pressing unresolved issue was whether Remicade treatment should be used solely to induce a short-term remission, perhaps as a bridge to longer-acting therapeutic agents. Despite the many open questions, the committee concluded the day-long hearing with the unanimous recommendation to approve Remicade for two specific indications: as a single-dose therapy for the treatment of Crohn's patients with moderate-to-severe disease symptoms resistant to conventional therapy, and as a three-dose infusion treatment for patients with actively draining enterocutaneous fistulae.

Recommendations of the advisory committees are usually, though not always, followed by the FDA. (One case in which the FDA did not follow a favorable recommendation of its advisory committee was the Centoxin application in 1992.) In the case of Remicade, the FDA complied with the recommendation and, on August 24, 1998, issued its approval for the two indications approved by the committee. It took another year to get approval for multidose therapy for all Crohn's patients with moderate-to-severe disease.

Remicade was the seventh therapeutic monoclonal antibody approved by the FDA and the very first one licensed for use in an inflammatory autoimmune disorder. (By March 2015 the number of approved therapeutic antibodies had climbed to around forty.) Approval of Remicade was granted on condition that postmarket monitoring would be undertaken to verify therapeutic benefit.

The approval came nearly ten years after the original mouse monoclonal antibody A2 had been generated in my NYU laboratory. Our wait time was not exceptional; a ten-year lag from the time a new therapeutic agent is generated to its regulatory approval is considered average.

Although Centocor had made the decision to seek approval for Crohn's disease before launching an effort to get Remicade licensed for rheumatoid arthritis, by 1995 the company had garnered sufficient resources to initiate a controlled clinical trial with Remicade (then still called cA2) in rheumatoid arthritis.

Before the end of 1999 the FDA granted approval for the use of Remicade in a three-dose infusion, administered in combination with methotrexate, in rheumatoid arthritis patients who had an inadequate response to methotrexate alone. Methotrexate, a synthetic drug originally used only for treatment of cancer, is a mainstay in the treatment of rheumatoid arthritis. Experience had shown that in rheumatoid arthritis Remicade was more effective when combined with methotrexate, partly because methotrexate may help to ward off a person's immune response to the administration of Remicade and partly because the two drugs act in concert.

Chapter Three

....

Miracle Drug

At NYU, my colleagues and I were following the successful approval process of Remicade for Crohn's disease and rheumatoid arthritis with personal satisfaction. The results of our work were helping to alleviate suffering from these diseases. Yet we didn't pop bottles of expensive champagne. Our reaction was muted, because in 1998–99 it was still far from clear that Remicade would have a very significant impact on the diseases for which it had been licensed.

In both Crohn's disease and in rheumatoid arthritis the treatment was approved only for short-term applications. Remicade did not cure the underlying disease and the symptoms returned when treatment was discontinued. There was only limited evidence suggesting a favorable therapeutic response when Remicade injections were repeated—perhaps three or four times—but there was not enough experience to know if long-term treatment was feasible. In principle, once a drug is approved, a physician may prescribe it essentially without restrictions, but in practice most physicians are reluctant to deviate from FDA recommendations, for liability reasons.

Questions remained: Would long-term administration of Remicade damage a person's immune system? Would defenses against infectious agents (bacteria, viruses, fungi, and parasitic agents) be too severely compromised? There was a plethora of animal data indicating that blocking TNF action weakens the body's defenses to infectious agents, such as *Mycobacterium tuberculosis*, the causative agent of tuberculosis. Also, would the body's immune response against the chimeric antibody protein constituting Remicade render repeated administration impossible? It was well documented that the human body may produce an immune response to the administration of chimeric antibodies.

No one had answers to these questions and my own bias during this time was that, more likely than not, these issues would preclude the wider use of Remicade. If long-term administration wasn't possible, the benefit of the treatment would amount to nothing more than short-term relief from the symptoms of two miserable diseases. Knowing that patients' disease symptoms are likely to relapse would almost certainly discourage physicians from prescribing the drug.

There were other obstacles. The treatment was expensive—even short-term administration would cost several thousand dollars. Remicade had to be administered by slow intravenous infusion in a specialized medical facility, usually inside a hospital or a specially equipped outpatient clinic. Finally, as a monoclonal antibody directed against TNF, Remicade embodied a radically new treatment concept, and, as with most new therapies, medical practitioners were not likely to rush to embrace it. Not surprisingly, initial Remicade sales were slow.

But there was one piece of news suggesting that not everyone was bearish about the prospects of Remicade: in 1999 Johnson & Johnson acquired Centocor for $4.9 billion. This seemed to be a very generous price given that Centocor's product sales were still modest.

In retrospect, Johnson & Johnson got a huge bargain. Subsequent clinical trials sponsored by Johnson & Johnson showed that Remicade was generally safe to administer to patients over a long term. In addition to Crohn's disease (including pediatric Crohn's disease) and rheumatoid arthritis, Remicade has also been approved—in the US and around the globe—for the use in ankylosing spondylitis (an inflammatory disease of the spine), ulceratitive colitis (another form of inflammatory bowel disease, including its pediatric form), juvenile idiopathic arthritis (a form of inflammatory arthritis occurring in children), psoriatic arthritis, and plaque psoriasis. It is estimated that through 2014 three million patients had been treated with Remicade worldwide.

The success of Remicade paved the way for the development and introduction of other drugs with a similar mechanism of action. Four other TNF-blocking agents are now approved for a variety of indications. (They are Amgen's Enbrel, AbbVie's Humira, UCB's Cimzia, and Johnson &

Johnson's Simponi.) It is fair to say that anti-TNF therapy has revolutionized the management of many chronic inflammatory disorders.

⸻

Most drugs are approved for one disease or perhaps a couple of closely
related diseases. Why is Remicade—along with the other anti-TNF
agents—effective in so many diseases that at first glance seem to have very
little in common? All of the diseases responsive to Remicade (and other
TNF-blocking agents) are autoimmune chronic inflammatory disorders,
in the pathogenesis of which TNF is now known to play a key role. In
autoimmunity—dubbed *horror autotoxicus* by Paul Ehrlich, a pioneer of
the study of autoimmunity—the immune system of the body attacks its
own tissues and organs.

The damage inflicted by autoimmune diseases results from the action of
antibodies and cells that constitute the immune system, the very same cells
that are important in protecting bodies from the attack by pathogens. The
result of such an attack is usually inflammation. In rheumatoid arthritis—a
prime example of an autoimmune disease—joints become swollen, warm,
and tender as a result of an inflammation of the soft tissue inside joints.

The site and extent of inflammation varies with each type of autoimmune disorder. Certain cytokines play important roles in initiating and
sustaining inflammation in autoimmune disorders, with TNF playing a
key role in some, but not all, autoimmune conditions. Because of its
important role in the inflammatory response accompanying many autoimmune disorders, curbing the action of TNF with Remicade or other
TNF-blocking treatments can be effective in numerous conditions.

Yet treatment with Remicade and other TNF-blocking agents does not
help all patients suffering from the diseases for which these treatments
are recommended. For reasons that are still poorly understood, around
30–40 percent of patients with rheumatoid arthritis, Crohn's disease, and
the other conditions for which Remicade is approved do not respond
adequately to these treatments. Unfortunately, it is still impossible to predict who will and who will not have a favorable response.

Moreover, Remicade and the other TNF-blocking agents can produce serious side effects in patients. Among the most common is an increased susceptibility to bacterial, viral, and fungal infections.

The treatments must be used with great caution in people who are infected with some pathogens even if they do not show overt signs of disease. For example, in people who have a dormant, symptomless infection with *Mycobacterium tuberculosis*, the administration of anti-TNF medications can lead to an activation of the infection resulting in overt tuberculosis. Appropriate precautions need to be taken with such patients. There is also a long-standing—though still unconfirmed—suspicion that in rare instances treatment with Remicade and other TNF-blocking agents might precipitate the occurrence of lymphomas and other malignancies, especially in children and adolescent patients.

And blocking TNF is not effective in all autoimmune diseases. In fact, in some autoimmune disorders, including multiple sclerosis and systemic lupus erythematosus (a common autoimmune disorder of women affecting the skin and multiple organs), Remicade and other anti-TNF agents are not only ineffective, they may make the disease worse. So these treatments can be used only in patients in whom the benefit from successful treatment outweighs the potential risks.

Another drawback is that anti-TNF medications, though bringing significant relief to patients who respond to the therapies, do not cure the underlying disease. If the treatment is stopped, patients relapse and the disease flares up anew. There are exceptions—I have met patients with Crohn's disease who after responding to Remicade treatment had gone off therapy and yet their disease failed to relapse for reasons that are not understood.

Thus autoimmune diseases responsive to treatment with anti-TNF agents have joined the list of many other chronic diseases for which there exist treatments providing powerful relief from the symptoms of the disease without eliminating the root cause of the illness. Diseases that can be controlled by appropriate treatment, but not cured, include diabetes that can be managed by treatment with insulin and other drugs, and even AIDS that can be kept in check by the administration of antiviral drugs. A major form of diabetes, type 1 diabetes, though not known to be responsive to anti-TNF therapy, is an autoimmune disease. AIDS is caused by infection with a virus, HIV, that takes up permanent residence inside certain vital white blood cells—especially so-called helper T lymphocytes—gradually leading to their destruction.

That anti-TNF therapies don't cure the underlying autoimmune disease is inherent in their mechanism of action. All of these agents act by blocking the activity of TNF, a key element in the chain of events causing inflammation. Inflammation, in turn, is the cause of the pathology in most autoimmune diseases, and upon disabling TNF and suppressing inflammation, disease symptoms cease. However, TNF is not the root cause of the disease—something else is—and if the administration of the anti-TNF agent is halted, TNF production and action generally resume, which is the reason why disease symptoms flare up anew.

Why is there no cure available for autoimmune diseases? The main reason is that we don't fully understand the underlying causes of these disorders. Indications are that certain environmental factors, including infections with bacteria or viruses, along with genetic factors built into our own DNA, determine whether we develop autoimmunity. Much more work will be needed before these factors become better known. And even when we learn the underlying cause of these illnesses, it still may take a long time to develop a cure. For example, we do understand the cause of diabetes—it is either the failure of beta cells in the pancreas to produce enough insulin or the failure of other cells in the body to respond to its action—and yet, available medications, though effective, generally do not cure the disease. There is hope that, as a result of more research and progress in fields like stem cell research and gene therapy, the situation will change in the foreseeable future.

Despite the limitations, Remicade and other TNF-blocking agents can be highly beneficial. Even though these therapies do not eliminate the underlying disease, they do much more than alleviate the symptoms of disease.

Specifically, in inflammatory bowel disease (either Crohn's disease or ulcerative colitis), anti-TNF agents can induce a clinical remission, promote the healing of lesions, prevent the appearance of new lesions, and eliminate the use of potentially harmful corticosteroids or the need for drastic surgery, thus often restoring the ability of patients to live normal, productive lives. Prior to the introduction of the anti-TNF agents, some patients with ulcerative colitis or Crohn's disease required colostomy—the surgical removal of much of their large intestine (colon) and attachment of its remaining portion to the outside of the abdominal wall. Intestinal waste would then flow through the opening in the abdomen into a colostomy bag. Anti-TNF therapy has largely eliminated the need for this surgery. Another benefit of anti-TNF therapy is a reduction in the high risk of gastrointestinal cancers in Crohn's disease and ulcerative colitis patients that correlates with the duration of the disease symptoms and severity of inflammation.

Similarly, in conditions whose major characteristic is joint inflammation (rheumatoid arthritis, psoriatic arthritis, ankylosing spondylitis), anti-TNF agents can arrest the progression of structural damage and improve physical function, thereby preventing the onset of a more disabling illness. These are no small accomplishments.

I've had the pleasure of meeting many patients treated with Remicade, who tell me how dramatically their lives have improved as a result of the medication.

And then there are letters and e-mails from patients or relatives of patients—many of them. Here are excerpts from three recent e-mails:

> I'm nineteen years old now, but when I was around thirteen years old I was diagnosed with Crohn's disease. When I was first diagnosed, and for about three years after that, my life was turning into a nightmare. The treatments I underwent at first, such as steroids, never worked. Many times they left me feeling more sick than before. I was in constant pain, and losing weight so rapidly that my mother had come to the point where she was not going to send me back to school. When my GI doctor told us about Remi-

cade, it was our last hope, I was so sick. From the first twenty-four hours since I was hooked up to my IV and given Remicade, I was a completely new person. I cannot even begin to tell you what it feels like to be better after being sick for so long. It gives you a completely different outlook [toward] the world around you and your purpose in the world . . .

I was diagnosed with ulcerative colitis at age thirteen and before my treatment with Remicade my life was incredibly difficult dealing with this painful condition. I am now thirty-nine years old and this miracle drug continues to work after starting it in Nov. 2006. I haven't had any problems whatsoever in these six years. In fact I have two healthy children and took the drug during my pregnancies . . . I just wanted you to know . . .

Five years ago our son, then seventeen years old, was diagnosed with Crohn's disease. Today, after receiving Remicade for four years, he had a completely normal colonoscopy. While dealing with this has not been easy for him, he has no way of knowing what we—his parents—know, that Crohn's disease used to mean pain and embarrassment, frequent hospitalizations, and regular surgeries. Our son has had a very normal life; even though he has had to take a handful of pills daily and receive intravenous medication every six to eight weeks, he has been able, along with his father, to raise thousands of dollars for Crohn's research by riding his bike fifty miles every year. He graduates from college in two weeks, a healthy, vibrant, and creative young man . . . Our child has had a relatively pain-free, healthy life. That has made a world of difference . . .

These letters remind me of my own good fortune: to have had the opportunity to contribute to a scientific venture that has made a real difference in the lives of millions of people.

Remicade became a huge economic success for Centocor (which in 2011 changed its name to Janssen Biotech) and Johnson & Johnson. When Remicade was first approved in 1998, initial sales were relatively modest. By 2001, with Remicade on the market for its third full year, worldwide sales reached $850 million. I had thought that was a very respectable result and had not expected much further growth. Ten years later, in 2011, worldwide sales exceeded $7 billion. From the outset, Remicade sales ran ahead not only of my own expectations, but also of the pundits' forecasts. In 2013, according to *PMLiVE*, a pharmaceutical industry newsletter, Remicade was the second-highest-selling drug in the world, with sales of $10.1 billion.

Billions of dollars are spent annually on researching and developing drugs, but it is estimated that just one out of approximately every five to ten thousand compounds studied in preclinical trials is ever going to be approved by the FDA and find its way to market. And of the drugs that are approved, just a very select few achieve blockbuster status, defined as reaching more than $1 billion in annual sales.

In 2013 the three highest-selling drugs in the world were all TNF inhibitors, a class of therapeutics whose development and introduction was trailblazed by Remicade. The number one best-selling drug was Humira—another TNF-blocking monoclonal antibody, manufactured and marketed by AbbVie, a company that in 2011 split off Abbott Laboratories—with 2013 worldwide sales stated by *PMLiVE* to have reached $11.1 billion. Approved by the FDA in 2000, about two years after Remicade, Humira is almost identical to Remicade in its efficacy and therapeutic spectrum. Whereas Remicade is a chimeric human-mouse antibody protein, with around 30 percent of its sequence stemming from the original mouse antibody, Humira is a fully human protein.

I am not aware of evidence that Remicade is less efficacious or that there are more complications observed with Remicade than with Humira. However, some patients and doctors may prefer Humira's route of administration; it is applied by injection under the skin, usually self-administered by the patient, whereas Remicade must be given by slow intravenous infusion in a specialized facility.

The third-highest-selling drug in the world in 2013, according to *PMLiVE*'s "top pharma list," was Amgen's Enbrel, a synthetic antibody-like TNF inhibitor, with sales of $8.9 billion. Enbrel too is used for indications almost identical to those of Remicade. Enbrel was first approved for use in rheumatoid arthritis in November 1998, a few months after Remicade's approval for Crohn's disease. Another pharmaceutical newsletter, *FierceBiotech*, listed Enbrel as the second-highest-selling drug in the world in 2013, with Humira in first place and Remicade in third.

In 2014, Humira retained its first place on *PMLiVE*'s top pharma list, but Remicade was reduced to third place by a newcomer—Solvadi, the "one thousand dollar pill" against hepatitis C, manufactured by Gilead Science.

Remicade has turned out to be a boon for NYU and, yes, for its NYU-based inventors. According to the licensing agreement signed by NYU and Centocor in 1984, NYU is due a royalty based on quarterly world-wide sales of Remicade. NYU, in turn, pays a portion of the collected royalties to Jimmy Le and me. Royalty payments collected by NYU for Remicade sales are not publicly disclosed, but it is safe to assume that by now they will have exceeded $1 billion. The medical and commercial success of Remicade illustrates the value of university-based biomedical research for therapeutic advances and economic progress. (I hope policy-makers in Washington will read these lines.)

Naturally, no party goes on forever. Most foreign Remicade patents expired in February 2015, with US patents due to run out in 2018. When patents covering conventional medications run out, the products may be replaced with generic drugs.

Monoclonal antibodies and other biologicals, usually large protein molecules produced by methods involving recombinant DNA tech-nology, are more complex than conventional small-molecule drugs that can be defined by chemical formulas. Biologics such as Remicade are not only structurally more complicated, their characteristics are also more likely to be influenced by changes in the manufacturing process.

Regulatory agencies in the US and in Europe have developed guidelines for the approval process of generic biologic drugs, referred to as biosimilars. In order to be approved for marketing, biosimilars must be shown to be close in properties to the parent biological product as demonstrated through analytical, animal, and clinical studies. The main yardstick used for the approval of biosimilar biologic drugs is the demonstration that they are comparable to the parent drug in clinical efficacy and safety.

The first biosimilar monoclonal antibody recommended for approval by the European Medicines Agency in October 2013 was a biosimilar Remicade, developed under the name Inflectra by the Korean company Celltrion. Inflectra and another Remicade biosimilar are already distributed in some European countries and in Canada, where Remicade is no longer under patent protection. According to news reports Inflectra is priced at about 20 to 30 percent below the annual cost of $10,000–$20,000 per patient for branded Remicade sold by Johnson & Johnson or Merck & Co., Johnson & Johnson's commercial partner in Europe. Additional fees that may be charged for the administration of the drug are likely to be the same for all forms of Remicade treatment.

I hope that introduction of a biosimilar Remicade will indeed make the treatment I helped to develop less costly and accessible to more patients who can benefit from its use.

————

People, including some less experienced scientists, often believe that scientific breakthroughs proceed in a logical and straightforward way from the eureka moment, when the discovery is supposedly conceived, presumably in the bathtub or—in more modern times—in the shower, through the validation of the original idea and the confirmation of the significance of the discovery, straight to the Nobel Prize or whatever the appropriate accolades may be.

In reality, as pointed out by James Watson, Nobel Prize–winning codiscoverer of the structure of DNA, in his succinct book, *The Double Helix*, "science seldom proceeds in the straightforward logical manner imagined by outsiders."

Remicade/Infliximab Milestones

1988-89
Monoclonal antibody specific for human TNF (named "A$_2$") is generated from a laboratory mouse at NYU School of Medicine

1990-91
Team of scientists at biotechnology company Centocor converts A$_2$ antibody by genetic manipulation into a "chimeric," predominantly human, monoclonal antibody (named "cA$_2$")

1991
Preclinical testing of cA$_2$ antibody at Centocor and NYU School of Medicine in test tube experiments and laboratory animals shows cA$_2$ is highly potent and selective, potentially suitable for human use

1991-92
Clinical trial of cA$_2$ in patients with sepsis showed no significant benefit

1992
cA$_2$ used successfully in rheumatoid arthritis patients by M. Feldmann, R. Maini, and colleagues

1993
cA$_2$ used successfully in Crohn's disease patients by S. van Deventer and colleagues

1998
cA$_2$ (by then named Remicade/infliximab) approved by the Food and Drug Administration (FDA) for the treatment of Crohn's disease

1999
FDA approval for rheumatoid arthritis

2004
FDA approval for ankylosing spondylitis

2005
FDA approval for psoriatic arthritis

FDA approval for ulcerative colitis

2006
FDA approval for severe plaque psoriasis

2013
Remicade/infliximab is the second-highest-grossing drug in the world

2014
Around three million patients treated worldwide with Remicade/infliximab

The A2 antibody that eventually became cA2 and Remicade is an object lesson in how science does not proceed in a "straightforward logical manner." It was first conceived as a tool for the development of a diagnostic test, with some vague allusions to its possible worth in medical applications related to autoimmunity and malignant tumors. When, some years later, it was thought that it might be useful as a therapeutic agent, the intended disease target was sepsis. When the sepsis trial failed to show positive results, the trajectory changed once again as cA2 was found to provide relief to patients with severe rheumatoid arthritis. The ultimate outcome—a therapy with profound effectiveness in a variety of severe autoimmune disorders—is a far cry from the original idea.

Serendipity played a huge role in getting the project off the ground and in seeing it through to a successful conclusion. The project would not have materialized if I had not run into Michael Wall during the 1982 meeting at Haverford College or if I had not accepted his invitation to visit the newly opened Centocor laboratories. The project would not have happened if Jimmy Le—then freshly trained in monoclonal antibody technology—had not joined my group. Remicade would almost certainly not have become what it is today if Marc Feldmann had not been able to convince Centocor (with the help of Feldmann's former colleague, Jim Woody) to provide him and Tiny Maini with cA2 for the first clinical trial in rheumatoid arthritis. I could go on.

Equally important in the successful outcome was the productive collaboration we were able to establish between my laboratory at NYU and the scientists and management at Centocor. Collaborations between academia and industry often falter because of their different cultures and because scientists in general with their strong egos have difficulty sharing credit. We got along and enjoyed our close interactions with the enthusiastic and dedicated Centocor scientists. Michael Wall and Hubert Schoemaker cared deeply about Centocor's success in the world of business, but they were also interested in the science behind our joint project.

It is fortunate too that my lab recognized the importance of TNF at a time when only a small group of people even knew that such a thing existed. Personal contacts played a role. The fact that I was friends with

Lloyd Old very likely influenced my decision to look for TNF in the materials we generated when producing IFN-gamma in my laboratory. In hindsight, there is a logical path from my experiments with interferon to cA2, yet in the moment, the path was not always clear. Yes, a similar drug may well have been developed elsewhere, as a result of other coincidences and serendipities, but these are the twists and turns that produced Remicade.

Or perhaps it can be all condensed to a sentence from E. B. White's love letter to the city, *Here is New York*: "No one should come to New York to live unless he is willing to be lucky." Amen.

PART TWO | # A Tumultuous Childhood

Chapter Four

....

Early Life

My father told me that our ancestors came from Moravia, a region between Bohemia and Slovakia, where the language spoken is Czech, somewhat different from Slovak. Yet, until 1945, our name was "Wilcsek"—an unlikely combination of characters since *w*, common in German and Polish, is not used in Czech, Slovak, or Hungarian, whereas *cs* is the uniquely Hungarian spelling for *ch* (as in "cherry").

The identically pronounced name "Wilczek" or "von Wilczek" belongs to an old aristocratic Austrian family with Polish roots. The mansion Palais Wilczek, one of the family's residencies, stands in the immediate vicinity of the Imperial Palace in Vienna. Whenever I am in Vienna, I jokingly point out the Palais Wilczek as our ancestral residence.

We are, of course, not related to the von Wilczeks, but the similarity of the names has led to a family theory about the possible origin of our surname. Until the late eighteenth century, very few Jews had last names at all. The situation in the Habsburg monarchy changed under the enlightened rule of Emperor Joseph II (1780–90), who abolished many oppressive anti-Jewish laws. One of the new laws introduced by Joseph II stipulated that every Jew in the then-vast Austrian and Hungarian territories was required to "adopt a German surname and never change it in his or her lifetime." Some Jews took on the name of their town, some the name of their profession. Still others are said to have adopted the name of their employer. Is it possible that some of my ancestors worked for a branch of the von Wilczek family?

By the mid-nineteenth century, records show numerous Wilcsek families living in the region close to where my father was born. They were all Jewish and, based on information collected by one of my distant cousins who now lives in Sweden, they were all related. As far as I know, as a result of the Holocaust, emigration, and defection, there are no more offspring of the Wilcseks living in Slovakia or Hungary today.

When my father, Július, was born in south-central Slovakia—in the small village of Jedľové Kostoľany—Slovakia was an integral part of Hungary; and Hungary, in turn, was an autonomous kingdom within the Austro-Hungarian Empire, ruled by the Habsburgs.

At the time of my father's birth, in 1896, my paternal grandfather, Vilmos Wilcsek, and his wife, grandmother Júlia, owned a dry-goods store. My grandfather's profession was common among Jews living in the countryside of what was then the Slovak region of Hungary. Owning a shop, often with an affiliated pub, was the way many Jewish families chose to make a living. Slovakia in those days was rural and quite poor.

The Hungarian government tried hard to suppress the Slovak national identity. Except for elementary schooling, all education was in Hungarian or, less commonly, in German. Slovak was the language predominantly spoken among the peasants, while all official business was transacted in Hungarian or German. Most Jews in Slovakia spoke Hungarian or German at home, which, along with the fact that many were small-business owners, did not endear them to the local Slovak-speaking population. Yiddish was not spoken (or even well understood) among the Jews living in most of Slovakia or Hungary. Only in the easternmost part of Hungary (in a region then known as Ruthenia or Subcarpathian Rus, now part of the Ukraine) and in Galicia (now split between Poland and the Ukraine) was Yiddish the main language of communication among the local Jewish population.

In 1906, when my father was ten, my grandfather decided to pack up and move his family to Budapest, the capital of Hungary. My grandfather opened a business on the outskirts of Budapest, supplying coal and firewood. He owned the shop until his death from a natural cause in the early 1940s. After the end of World War II, with anti-German sentiments on the rise, my father changed the spelling of our family name from Wilcsek to Vilček, in tune with the Slovak orthography.

In Budapest, my father completed the equivalent of high school. He was not yet eighteen at the time World War I broke out in 1914. Within a year he was drafted into the Austro-Hungarian Army. Because he had completed secondary school he enlisted as *Einjährig Freiwilliger*, a volunteer for one year, a misnomer because he was neither a volunteer nor was he able to leave the army after one year.

He served on the Italian front until the end of the war in 1918—which marked the demise of the Austro-Hungarian Empire—but he rarely talked about the years spent in the army. I still have some photographs of him from those years in which he looks quite dashing in his smart officer's uniform. One story he told was how at Christmastime there would be armistice declared between the two warring armies, and he and his fellow Austro-Hungarian officers would have a good time with Italian officers behind enemy lines. Unlike others who lived through similar experiences, he never complained about the horrors of war or anything else. He was, like me, a perpetual optimist.

After the end of World War I, my father decided to move back to the land of his birth, Slovakia, which had by then separated from Hungary and become part of the newly formed Czechoslovak Republic. He settled in Bratislava—then the regional capital of Slovakia in the country of Czechoslovakia, and today the capital of an independent Slovak Republic.

Sometime in the 1920s my father joined a company called Handlovské Uhoľné Bane (Handlová Coal Mines). Although his education had gone no further than high school, by the time he married my mother in 1928 he held the position of *prokurista*, an executive authorized to sign official documents for the company. He was responsible for the sales of coal produced by the mines in the town of Handlová in central Slovakia, though he worked from the company's headquarters in Bratislava. His income enabled our family to live an upper-middle-class life. His office was within walking distance from our home, but still he was picked up in the morning and delivered back home in the evening by a chauffeured company car. The same chauffeured car would usually bring him home at lunchtime and return to pick him up again after the then-customary midday siesta.

My father was a happy, gregarious person. He would strike up friendly conversations with strangers—waiters, taxi drivers, or any other person who happened to cross his path. His rapid rise in the company was not due to his social standing or connections. He was completely fluent in Hungarian, Slovak, and German; and initially all three languages were of relatively equal importance in the conduct of his business. He was the quintessential self-made man, appreciated and promoted for his natural intelligence, business acumen, and loyalty.

———————

I know more about the history of my mother's family, mainly because she liked to talk about her relatives more than my father did. My great-grandparents, the Kleins, moved to Budapest from the Burgenland region of Austria, which borders Hungary. Because they came from Austria, the family continued to use German as their main language of communication.

My maternal grandmother, Júlia, married Alexander Fischer, a young employee of an Austrian bank in Budapest. Their only daughter, my mother, Friderika, nicknamed Fricy, was born in 1907. While growing up in Budapest my mother was particularly close to three of her cousins, all boys. In her quest to be considered the smartest child in the extended family, my mother enrolled in an academic high school, called a *gymnázium* in Central Europe, when she was eleven. Only academically inclined children were admitted to a gymnázium.

Then, sometime shortly after the end of World War I, my grandfather was transferred by his Austrian bank to Bratislava, and my mother, by then a teenager, moved with her parents to a new city and a new country. At the time my mother spoke no Slovak or Czech, the official languages of Czechoslovakia. Fortunately, Bratislava in those days was quite cosmopolitan and much of the local population was more comfortable speaking German or Hungarian than Slovak. (Until 1919, the city was known as Pozsony in Hungarian, Pressburg in German, and Presbourg in French.) My mother enrolled in a Hungarian gymnázium known for its high academic standard.

For the rest of her life, my mother would proudly talk about her studies of Latin and ancient Greek, along with French, English, and world literature. Hungarian always remained her preferred language, the language in which she was most secure and comfortable. She loved Hungarian literature, pointing out that if it weren't for the unfortunate fact that so few people understood the language, Hungarian novelists and poets would be considered the best in the world. Had she lived longer she would have been pleased to learn that Imre Kertész was awarded the Nobel Prize in Literature in 2002 and that the writings of Sándor Márai have become widely known all over the world.

My grandparents also made sure that their daughter would acquire proper social graces. During summers she was sent to girls' schools in Geneva and Lausanne, Switzerland, in order to perfect her command of languages and to learn manners appropriate for a young upper-middle-class lady of those days. Following high school, my mother enrolled in medical school in Bratislava. To succeed she had to perfect her Czech and Slovak. There were only three girls in her medical school class among over one hundred male students. She completed her degree in 1931, passed her medical license examination a year later, and became an ophthalmologist.

My parents married in 1928 when my mother was twenty-one and still a medical student; my father was nearly thirty-two. I am not sure why my mother was in such a hurry to get married. True, in those days women generally married very young, and even though my mother was an attractive and smart young woman, it is possible she worried that waiting until she became a doctor might impair her marriageability.

Curiously, my parents never talked to me about where and how they met. They also never mentioned their wedding and I have not seen any photographs of a wedding celebration. There should have been a Jewish wedding ceremony, but all I was able to find among my parents' documents was a marriage certificate issued by a municipal agency in Bratislava confirming that they were married "in front of" a city official. Was it an arranged marriage? Was my mother perhaps under pressure to marry someone affluent to be able to complete her medical studies? I remember her making vague references to financial hardship experienced by her parents during the Great Depression. Whatever the reasons, I am quite sure she was not compelled to marry by an unwanted pregnancy, as I, her only son, was born five years later.

My mother often hinted that her marriage was a *mésalliance* because she was better educated and because my father's family was inferior in social standing and sophistication to hers. They seemed to get along fine in the early 1930s, a period during which they traveled widely in Europe together. I found letters my father had written to my mother in which he addresses her *Drága Mukikém* ("My dear Muki"), a tender Hungarian nickname I never heard him use during my lifetime.

Apart from their frequent arguments and mutual criticism, my parents lived a comfortable life. My father continued to work for Handlovské Uhoľné Bane. In the mid-1930s, after finishing her training in ophthalmology at the university clinic in Bratislava (with stints at universities in Vienna and London), my mother opened a small private practice in our apartment. Having the practice at home was convenient, especially after I was born in 1933. We lived at the time on the centrally located Grösslingova Street, in a nice, spacious, but not overly luxurious rental apartment, where my mother used one room as her professional doctor's office. I don't think the practice ever became lucrative; my father earned enough money to support the family, so it was more a matter of my mother's pride that she continued to practice ophthalmology even though she didn't really have to work.

As I got older, I became aware of the fact that over the many years they lived together, my mother had some boyfriends and my father had girlfriends. Extramarital affairs were quite common in their social circle, and not considered as grave a betrayal as in more recent times. Divorces in those days were extremely rare, and, to make an unsuccessful marriage tolerable, couples of my parents' social class sometimes would make "arrangements." There was, in fact, a well-established institution of "house friends," men who more or less openly "entertained" wives of unloved or unloving husbands. Not unlike the story of the Marschallin and Octavian in Richard Strauss's *Rosenkavalier*, this kind of arrangement had probably been in existence for centuries.

My mother's house friend in the late 1930s was a professional colleague and friend of my father named Otto. He was very kind to me. As a child I was unable to properly pronounce his first name and used to call him *Onkel Motto*. It was common in those days that children addressed friends of their parents as "uncle" or "aunt" even if there was no family relationship. At the time I didn't fully understand the relationship he had with my mother. It wasn't until after the war that I realized neither of my parents was faithful.

Onkel Motto died in Auschwitz during the Holocaust. I remember that my mother was heartbroken when she learned the tragic news. For many years she had kept a framed photograph of Onkel Motto at her bedside—in full view of my father.

The relationship between my mother and father visibly improved in the last two decades of their lives, when they became more dependent on each other's support. In 1979, after over fifty years of marriage, my parents died within months of each other: my father first, of stomach cancer, which he had had for several years; then, only a few months later, my mother, who had otherwise been in good health, broke her hip and died suddenly of a pulmonary embolism.

My parents had a complicated attitude toward their Jewishness. In their day, religious affiliations were always recorded in official documents, including birth certificates, identification cards, and marriage certificates. My parents' documents listed their religions as *izr.*, a Slovak abbreviation for "Israelite." Throughout their lives, most personal friends of my parents were Jewish. Both my father and my mother had always been proud of their Jewish heritage, often pointing out to me the disproportionately large number of Jewish Nobel Prize winners, famous artists, and other outstanding individuals. They would refer to fellow Jews as "our people" (they either used the term *našinec* in Slovak, or *Unsereiner* in German). After the German annexation (*Anschluss*) of Austria in March 1938, my parents volunteered to help Austrian Jews who sought refuge in Bratislava, then still part of a democratic, free Czechoslovakia. Yet my parents—especially my mother—were generally reluctant to wear Jewishness on their sleeves, particularly in front of gentiles, not only during the Second World War when being identified as a Jew often equaled a death sentence, but also before that and long thereafter. For them Jewishness was something acknowledged among family members and Jewish friends, but it was generally not to be discussed with others.

I remember that I once talked about our Jewish heritage with my non-Jewish friends in the presence of my mother. When my friends left she admonished me and strongly advised me not to talk about my origin in public. My father was more open-minded; even though he did not speak Yiddish, he occasionally used expressions like *nebbish* or *shlemiel.* I never knew my parents to be members of any Jewish congregation, nor was I aware of them ever visiting a synagogue. Jewish holidays were not celebrated and in fact not even acknowledged in our house.

In contrast, we had a tree at home at Christmastime and I would eagerly await my presents on Christmas Eve. In this respect my parents were not unique; most of their Jewish friends were not observant. In fact, even before the rise of Nazism, many Jews in Austria, Germany, Hungary, and Czechoslovakia—especially upper-class, educated Jews—were quite lax or not observant at all. Being Jewish, and especially being an openly observant Jew, was not helpful for professional advancement. My mother would sometimes make mildly critical comments about Jewish customs she did not appreciate. When I attempted to move food from my plate onto hers or from hers onto mine she would admonish me, "Don't behave like guests at a Jewish wedding."

And so I was brought up with no religious faith or affiliation, until 1939, when right before my sixth birthday my parents decided to have me christened and my mother converted to Catholicism. By then, of course, the persecution of Jews was in full swing in Germany and Austria, Czechoslovakia had been crushed and dismantled as a result of the occupation of the Czech lands by Hitler's army, and Slovakia had become a satellite state of Nazi Germany, governed by a nationalist Fascist government. With the support of some members of the Catholic clergy, Slovakia was about to introduce the equivalent of the Nuremberg Laws severely restricting Jewish rights and freedoms. My father converted to Christianity more reluctantly three years later, in the spring of 1942.

By converting to Catholicism my parents hoped to avoid the otherwise inevitable consequences of the impending persecution of the Jewish population. Although they were not able to escape persecution, in retrospect it is clear that their formal acceptance of the mantle of Christianity almost certainly saved all three of our lives.

———

My favorite companion in early childhood was our dog Kalos, meaning "beautiful" in ancient Greek, a white miniature poodle about one year older than me. He was given his name by my mother, who, I suspect, wanted to show off her command of ancient Greek. Apparently, Kalos did not resent the attention I stole from him when I was born. I loved him dearly, and he requited my love. We appear together in dozens of photographs, my favorite being one in which I am pushing a brightly smiling Kalos in my own baby carriage. (If you have doubts whether dogs are capable of smiling, look at the photo.)

In the first three years of my life, my mother kept combined diary-photo albums, one for each year. My mother's comments are written in German, in a neat but not easily legible cursive handwriting. It is amazing that the diaries survived the demise of the first Czechoslovak Republic, the upheavals around the outbreak of the Second World War, the Holocaust, reestablishment of free Czechoslovakia after the war, the Communist takeover, and the years of Communist rule, not to mention the umpteen voluntary and forced moves my parents made during all those years.

My main caretaker until I was about a year and a half was a professionally trained nanny, Hildegard—a German-speaking Austrian woman. She was officially addressed as *Schwester* (sister), a contraction of the German term *Kinderschwester* or nanny. In photographs I've seen, Hilde is a big, intimidating woman, dressed in a uniform that makes her look like a Protestant nun. This must have been the dress code for nannies in those days.

My next nanny was a Slovak woman whose name, I think, was Terka, a contraction of Tereza. Until that moment in time the only language spoken to me was German, but Terka was not a German speaker, and she communicated with me in Slovak, a language I picked up easily, as children often do.

The third language I was exposed to early on was Hungarian. I don't think anyone in the family spoke to me in Hungarian, but Hungarian was the principal language of communication between my parents (especially when they did not want me to understand what they were saying), and also among the friends of my parents who paid frequent visits to our house. My mother and I, along with the nanny who took care of me at the time, spent the whole summers following my first and second birthdays at Balatonlelle, a popular resort town on the southern shore of Lake Balaton in Hungary. I was too young to retain any memories of these summer holidays, but while there I must have played with children who spoke Hungarian, and most of the adults around would likely have spoken Hungarian too.

By age four or five I was able to communicate in three languages with almost equal ease. Curiously, although my German and especially my Hungarian are by no means flawless, I have retained to this day the ability to converse in those three languages. I spent a great deal more effort later in life to study Russian, French, and English.

———

Adolf Hitler was appointed chancellor of the German Reich in January 1933—about five months before my birth. Almost immediately, the Nazi government started to restrict professional activities of its Jewish citizens.

By 1935, German authorities adopted the Nuremberg Race Laws, which defined Jews by their ancestral lineage rather than their religious affiliation. These actions were followed by severe restrictions of economic activities, including "Aryanization" (meaning expropriation of Jewish-owned businesses and their transfer to non-Jewish owners), essentially preventing most Jews in Germany from making a living.

Like people in other countries surrounding Germany, my parents watched these developments with anxiety. However, many, including a significant proportion of German Jews, thought that these were temporary excesses that would soon come to an end. My parents often repeated the modified lyrics of a then-trendy pop song—*Es geht alles vorüber, es geht alles vorbei, erst geht der Führer und dann die Partei* (It will all pass, it will all go, first goes the führer, and then his party)—a spoof of the lyrics of a pro-militaristic song originally recorded by German singer Lale Andersen, who had also popularized "Lili Marleen" before Marlene Dietrich's version made it a hit among Allied soldiers during the Second World War.

Czechoslovakia at the time was still a free democratic state, and my family continued, at first, to live a normal, comfortable life. Most Czechoslovak Jews, including my parents, while distressed about the situation in Germany, believed that Hitler's rule would soon end and that, in any case, Czechoslovakia, with her strong army and her European allies would be able to defend herself against any foolish attempt by Nazi Germany to interfere with her democratic system.

This generally optimistic, even careless attitude toward the upheavals in Nazi Germany ended abruptly in 1938 as a result of two events. One was the Austrian Anschluss, completed in March by the transfer of sovereign Austrian power to Germany and the triumphal march of Hitler alongside his Wehrmacht troops into Vienna. My mother happened to be in Vienna at the time, and she witnessed the huge crowds of Austrians beside themselves with joy welcoming the führer and cheering the Anschluss. "Until that day, Austria felt like my second home," I remember my mother saying. "I will never forget the insanity of these people."

Immediately following the Anschluss, a hounding of Jews got underway

Some Major Events Affecting Territories of Former Czechoslovakia 1918–2004

1918	October 28	Czechoslovakia established after the defeat of Austro-Hungarian Empire in World War I
1933	January 30	Hitler becomes chancellor of Germany
1938	March 12	Anschluss: Austria annexed by Germany
	September 30	Signing of Munich Agreement, forcing Czechoslovakia to yield the Sudetenland border region to Germany
	October 5	Edvard Beneš resigns as president of Czechoslovakia
	October 10	Sudetenland annexed by Germany
	November 2	Southern Slovakia and a segment of Subcarpathian Rus (also known as Ruthenia, now part of the Ukraine) annexed by Hungary
	November 19	In response to demands by Slovak nationalists, Czechoslovakia's name is changed to Czecho-Slovakia and Slovakia is granted autonomy
1939	March 14	Under German pressure Slovakia secedes from Czechoslovakia; establishment of Slovak State governed by Fascist government allied with Nazi Germany
	March 15	Germany occupies the rest of Czechoslovakia; Hitler announces creation of the Protectorate of Bohemia and Moravia
	April 18	Slovak lawmakers pass first anti-Jewish laws
	September 1	Germany invades Poland; two days later Britain, France, Australia, and New Zealand declare war on Germany

	October 26	Jozef Tiso elected president and Vojtech Tuka prime minister of the Slovak State
1940	April 25	Slovak lawmakers pass the first Aryanization law
1941	September 9	Slovak government introduces Jewish Codex (which is a more drastic version of the German Nuremberg Laws)
	October 23	Germany pledges help to Slovak representatives with the solution of the "Jewish problem"; SS leader Himmler announces creation of dedicated areas for European Jews in Poland
	December 2	Prime Minister Tuka negotiates with German Ambassador Hanns Ludin details of forced Jewish deportations; Tuka agrees to pay Germany 500 marks for each deported person as a "resettlement fee"
	December 11	United States enters war against Germany
1942	March–October	First wave of Jewish deportations from Slovakia: fifty-eight thousand (approximately two-thirds of the entire Jewish population of Slovakia) are transported to concentration and extermination camps
1943	February 2	Germany loses the battle of Stalingrad
1944	August 29	German troops occupy Slovakia; start of the Slovak National Uprising
	October 27	Slovak National Uprising crushed (though guerrilla attacks continue until the end of the war)
1945	March 31	Last transport of Jews from Slovakia to concentration camps
	May 9	End of World War II in Europe

1948	February 25	Communist coup ends democratic system of government in postwar Czechoslovakia
1953	March 5	Death of Stalin
1956	February 25	Khrushchev delivers secret speech criticizing Stalin's brutal rule at the Twentieth Congress of the Communist Party of the Soviet Union
1962	October	Cuban Missile Crisis
1968	January–August	"Prague Spring" or "socialism with a human face" introduced in Czechoslovakia by the reform-minded Communist leader Alexander Dubček
	August 21	Five Warsaw Pact armies under Soviet command invade Czechoslovakia to crush the democratization process. Soon the process of "normalization" begins in Czechoslovakia under the new Communist Party leader and later president Gustav Husák
1989	November 17	"Velvet Revolution" against Communist system begins in Czechoslovakia
	December 29	Václav Havel is democratically elected president of Czechoslovakia
1993	January 1	Czechoslovakia is peacefully dissolved into the Czech and Slovak Republics ("Velvet Divorce")
1999	March 12	Czech Republic joins NATO
2004	March 29	Slovak Republic joins NATO
	May 1	Czech and Slovak Republics become members of the European Union

in Austria that was even more brutal in nature than the persecutions in Germany. With only forty miles between Vienna and Bratislava, the events in Austria suddenly became a grave concern for my family and their friends. What began as restrictions on the life and liberties of Austrian Jews rapidly progressed toward the extermination of Jews unable or unwilling to emigrate within a relatively narrow time frame. It is estimated that in 1938, Austria had a Jewish population of nearly 200,000, largely centered in Vienna. Some 120,000 were forced to emigrate, most often leaving behind all of their possessions. The number of Austrian Jews killed in the Holocaust is estimated to be 65,000.

The second set of events that cast a long shadow over the security of Jews in Czechoslovakia were the developments that culminated in the signing of the Munich Agreement by Germany, France, the United Kingdom, and Italy on September 30, 1938. Forced upon the Czechoslovak government by Neville Chamberlain of Great Britain and Édouard Daladier of France, the agreement authorized Germany's annexation of Sudetenland, Czech territories along the country's northern, western, and southern borders, inhabited by some three million German speakers. It is referred to as the Munich betrayal by the Czechs and Slovaks, and today it is widely regarded as a failed attempt at appeasing Nazi Germany.

Sudetenland was of immense strategic importance to Czechoslovakia, as most of her border defenses and heavy industries were situated there. Without the annexed territories, Czechoslovakia no longer had a viable chance to defend herself against a German invasion. It was now clear that the population of Czechoslovakia, including her Jewish citizens, could no longer count on the Czechoslovak army and its Western European allies to protect the democratic system in the country.

Only a few weeks after the signing of the Munich Agreement, another territorial concession was forced upon Czechoslovakia by Nazi Germany and Fascist Italy, allowing Hungary to annex territories in Southern Slovakia and part of the easternmost region of Subcarpathian Rus, populated predominantly by ethnic Hungarians. The territory annexed by Hungary included Košice, the second-largest city in Slovakia.

In June of 1938, midway between the time of the Anschluss and the signing of the Munich Agreement, my mother took me on a trip to Amsterdam. In order to avoid crossing Nazi Germany into Holland, we were traveling from Prague to Amsterdam on a commercial flight, at the time still a novel and unusual form of transportation. It was the first time that I traveled on an airplane; it was likely my mother's first flight as well, because my parents' earlier European travels were all by train. They didn't own or drive a car.

Exciting as flying to a new destination was, this was not strictly a pleasure trip. After the Austrian Anschluss, the distance from our home in the center of Bratislava to the border with Nazi Germany was only a few miles. And even though the Munich Agreement had not yet been signed, Hitler's threatening rhetoric had made it very clear that he was ready to move aggressively against Czechoslovakia.

With this sword of Damocles dangling over their heads, my parents started to consider emigration to a country where they would be safely out of Nazi Germany's reach. At the time, Holland seemed a good choice. Hitler was outspoken about his aspirations to liberate "oppressed" German minorities in Czechoslovakia and Poland, but there was no indication that he would consider moving against countries beyond the western borders of Germany. And if he were to try, France and England would certainly stop him. There was also a personal reason for considering Holland as a potential safe haven. Some years earlier, while vacationing at a seaside Adriatic resort in Italy, my parents befriended a Dutch Jewish couple, Joost and Juliette Egger. They too had only one child, Selma, a few months my senior. Jo was a successful importer and film producer.

I was about to turn five and, naturally, knew nothing about the main purpose of our trip. I was told that we were going to visit our friends, Uncle Jo and his wife Juliette, and that I would get to know Selma. Indeed, we stayed in their townhouse in central Amsterdam, at Den Texstraat 5, which today comprises a small hotel. I remember our visit to the Amsterdam zoo, especially an alley lined with cages of parakeets that greeted everyone with a loud *goedendag*, "good day" in Dutch. Our Dutch hosts also took us to a seaside resort, Scheveningen. It was there that I celebrated my fifth birthday.

Unbeknownst to me at the time, my mother was carrying with her some of the most valuable pieces of our family jewelry. Before our departure from Amsterdam she left the jewelry in safekeeping with the Eggers. This, together with whatever cash my parents would be able to carry when fleeing Czechoslovakia, was going to tide us over for the first weeks or months after emigration. There was another plan my mother revealed to me only when I was an adult. She was considering leaving me in the house of her Dutch friends because, she believed, I would be much safer in Holland than in Czechoslovakia, where a Nazi takeover seemed imminent. The plan was that my parents would join me soon thereafter. Fortunately, given how history developed, I did return to Bratislava with my mother; and my parents' plan to emigrate to Holland never amounted to more than a pipe dream.

Less than two years after our visit to Amsterdam, Nazi Germany invaded Holland; a few days later the Dutch forces surrendered. In May 1945, my mother tried to get in touch with our friends, but her letters and inquiries went unanswered. In the fall of 1945, my father decided to take the long trip through war-ravaged Europe to Amsterdam in order to find out what had happened. He found the Eggers' house in Amsterdam, only to learn from the neighbors that Jo, Juliette, and Selma were rounded up when the deportations of Jews in the Netherlands got underway in 1942. They all perished. Miraculously, one of their Christian neighbors returned to my father a golden cigarette case with a ruby set into its clasp. The neighbors told my father that, before they had been rounded up, our friends entrusted this piece to them, explaining that it belonged to a family in Czechoslovakia. None of the many other pieces of jewelry my mother took to Holland have ever resurfaced.

Most of the silver owned by our family was lost too. Shortly after the establishment of the Fascist Slovak State my father was arrested and kept in jail for several days without ever being told why. Being Jewish was probably enough of a reason. When my mother went to negotiate for his release, the police official agreed to let him go—in exchange for the family silver.

When Jewish persecution started in Slovakia after the breakup of Czechoslovakia, Jews were ordered to deposit all their gold jewelry and other precious objects with the authorities. Not expecting to ever see the deposited valuables again, my parents surrendered only some items, hiding the rest with acquaintances. Like the jewelry left in Holland, most of the valuables hidden in Slovakia were never returned. Yet, after the end of the war, my parents did recover every single item of jewelry that had been surrendered to the Slovak Fascist government authorities.

Chapter Five

War

Hitler was not appeased by the gains he made as a result of the Munich Agreement and, less than six months after the annexation of the Sudetenland region, he moved to crush whatever was left of Czechoslovakia. On March 15, 1939, the Wehrmacht occupied the Czech lands, establishing the "Protectorate of Bohemia and Moravia." Though not formally annexed by Germany, Bohemia and Moravia became for all practical purposes a country under German rule. At about the same time Hitler gave Hungary permission to occupy the part of the region of Subcarpathian Rus that until then had formally remained part of Czechoslovakia.

For Slovakia, Hitler had different designs. With the cooperation of the right-wing Slovak Nationalist Party under the leadership of the Catholic priest Jozef Tiso, an independent Slovak State was established on March 14, with Tiso becoming its president. Deputies who were initially reluctant to ratify the new Slovak State were "persuaded" when Germany issued a threat that Slovakia would be partitioned between Hungary and Poland. Tiso then sent Hitler a telegram, drafted for him by the Germans, asking that Germany kindly take over the protection of the newly established state.

The new Slovak government wasted no time in starting to curtail the rights of the Jewish population. In April 1939, barely a month after the establishment of the Fascist Slovak State, the new parliament adopted a law limiting the participation of Jews in certain professions to 4 percent. This action was followed by a series of laws prohibiting Jews from owning businesses and introducing the concept of "Aryanization," leading to the replacement of Jewish owners by new Aryan owners.

In September 1941, the Slovak government passed the so-called Jewish

Codex, a comprehensive set of regulations stripping Jews—defined on the basis of "race" and ancestry, not religious observance—of all basic rights and liberties. In addition to prohibiting Jews from running for public office, voting, carrying out any professional activities, holding jobs in government or public organizations, and attending school beyond the elementary school level, the sixty-page document went as far as specifically prohibiting Jews from fishing, driving, owning cameras, and carrying or having at home more than the equivalent of forty dollars in cash. So that they would be readily identifiable, Jews were ordered to affix on their outer garment at all times a clearly visible yellow Star of David.

In 1938, a few months before the dissolution of Czechoslovakia, our family had moved to an apartment in a newly constructed building in the center of Bratislava. Like our previous apartment, it too was a rental. My parents never owned a house or an apartment in their lives. In the late 1930s they had bought a plot in a prime spot on a hill overlooking Bratislava and the Danube River, and were planning to build a home on it, but their plans were thwarted by the establishment of the Slovak State. The land was Aryanized, only to be returned to us after the end of the war, but then the advent of Communism once again interfered with the construction of a family residence.

The new apartment in the center of Bratislava was not roomier than the old one, but it came with amenities such as built-in closets, which were still quite rare in those days. We did not get to enjoy living in the new apartment for very long. One of the provisions of the discriminatory laws was that Jews were not allowed to live in certain sections of towns. We were ordered to move to a much smaller, inferior apartment in a working-class neighborhood of Bratislava called Tehelné pole ("Brick Field").

We were among many Jewish families ordered to move to that section of town, to the dismay of a portion of the original non-Jewish population. A large wooden fence in our neighborhood was decorated with a freshly painted message, "JEWS, THE ROAD TO PALESTINE DOES NOT LEAD THROUGH TEHELNÉ POLE!" Until our move to Tehelné pole I had not experienced overt anti-Semitism. Soon after our move, as I bicycled around the block that included our new home, a group of neighborhood children and adults formed a human wall to stop me. Forced to dismount my bicycle, they screamed at me to get out of their neighborhood and go bicycle to Palestine instead.

Passing the Jewish Codex enabled the government to make life utterly miserable for Jews, but our lives were not yet threatened. This was to change very soon. In December 1941, the Slovak premier Vojtech Tuka, a member of the radical pro-Nazi wing of the government, negotiated an agreement with the German ambassador to Slovakia, Hanns E. Ludin, which authorized the forced deportation of Jews from Slovakia to territories controlled by Germany. So eager were the Slovak Nazis to get rid of the Jews, they even agreed to pay a "resettling fee" of five hundred German marks for every Jew accepted by the German authorities. By October 1942, some fifty-eight thousand Jews—two-thirds of all Jews living in Slovakia—were forcibly deported to German concentration camps, mainly Auschwitz; very few of the deportees survived to the end of the war.

Up to this point the persecution of Jews in Slovakia followed a script very similar to that being applied in Germany, Austria, the Czech lands, and Poland. Eradication of Jews in the Baltic countries and parts of the Soviet Union occupied by German troops in 1941–42 was even more violent and brutal. In contrast to these countries and regions under direct German control, deportations of Jews from Slovakia were suddenly halted in October 1942.

According to some sources, the cessation of deportations happened as a result of efforts of surviving Jewish leaders, who made sure that members of the Slovak government clearly understood that deported Jews were not being "resettled" but slated for extermination. It was also the result of help from more moderate members of the Slovak pro-Nazi government and perhaps a slight push from Vatican officials interested mainly in protecting Jews who had converted to Catholicism. The Jews who were left in Slovakia after the cessation of deportations were still subject to the harsh treatment mandated by the discriminatory laws, but for nearly two years there were no mass deportations to concentration camps in Germany or German-occupied territories.

Changes in the Boundaries of Czechoslovakia 1918–93

CZECHOSLOVAKIA 1918 – 1938

**Partition of
CZECHOSLOVAKIA 1938 – 1939**

CZECHOSLOVAKIA 1945 – 1993

CZECH REPUBLIC and SLOVAKIA
1993 – present

Another Slovak anomaly was that some Jews were able to obtain legal exemptions from persecutions imposed by the Jewish Codex. One type of exemption that could be granted only by Jozef Tiso constituted a partial or complete pardon from provisions of the Jewish Codex. People awarded a complete pardon were for most practical purposes once again treated as ordinary citizens. Other types of partial exemptions could be granted by several branches of the Slovak government. Jews who were able to obtain such exemptions (called *výnimky*) were allowed to practice their professions, albeit with restrictions, and they were not required to display yellow Stars of David on their garments.

The *výnimky* granting process was opaque, and as far as I know, exact records of how they were granted and how many were issued have not been preserved. What is known is that hefty administrative fees had to be paid for the exemptions, and an estimated total of five to ten thousand individuals benefitted from them. Persons granted these exemptions tended to be affluent, educated, and, almost without exception, converts to Christianity.

The official justification for granting these exemptions was that, at least in the short run, Slovakia could not function without educated Jews, because there weren't enough Aryan physicians, dentists, pharmacists, people with engineering degrees, or economists. Connections and bribes likely played a significant role, too, in the decision of who did and who did not receive what in reality amounted to a reprieve from a death sentence. The laws made clear that *výnimky* could be withdrawn at any time.

In 1942 my parents applied for and were granted exemptions from the provisions of the Jewish Codex. On behalf of herself and me, my mother was granted a full exemption from all provisions of the Codex, probably because she was a physician, and she had converted to Catholicism at an early date. My father was considered essential for the Slovak wartime economy in view of his intimate understanding of the national coal business, but he received only a partial exemption, perhaps because he converted to Catholicism later than my mother. His partial exemption meant that he did not have to wear the yellow star and he was allowed to work, but he was still prohibited from owning a camera, going fishing, and driving a car, along with numerous other restrictions. My father was an avid photographer, but fortunately he was not fond of fishing and he had never learned how to drive.

———

I completed the first two grades of elementary school in a public boys' school located near our family home on Grösslingova Street. In a group photo from the first grade I can still identify two Jewish boys among my thirty-nine classmates. By the time I was ready to advance to the third grade in the fall of 1941, my parents knew that I would no longer be allowed to continue attending a regular public school. They could have enrolled me in a specially designated school for Jewish children, but my parents decided not to do that.

To protect me from the possibility that I would be sent with them to a concentration camp, my parents chose an unconventional path: they placed me in an orphanage with an affiliated school run by Catholic nuns. The nuns belonged to the Daughters of Charity of St. Vincent de Paul. Recognizable by their huge triangular starched white hats, the St. Vincent's Sisters were known for devoting their lives to the education of needy children.

The orphanage and school occupied an imposing Victorian building on Hlboká Street, to this day a prominent and well-known structure in Bratislava. Even though I was by then officially a christened Catholic, accepting me was almost certainly illegal at the time and it was brave of the St. Vincent's Sisters to do it. I entered the school with another Jewish boy of the same age, my close friend and classmate from the first and second grades, Jan Deutsch.

When I mention that at age eight I was separated from my parents and placed in an orphanage, people assume that it must have been a traumatic experience. Some memories I retain are certainly Dickensian. I slept in a huge room with about fifty other boys—most of them orphans. The boys who wetted their sheets during the night (I was not among them) were required to stand in the morning with their soiled sheets in hand for all to see. Supervising us during the times we were not in class, especially in the dining room and during sports, was a man whose official title was "prefect" (we addressed him *pán prefekt*, literally "Mr. Prefect") who walked around with a thin bamboo stick, ready to hit any boy who misbehaved. But most of my other memories are benign. I recall that the St. Vincent Sisters were devoted teachers and I liked them. There was heavy emphasis on Catholic religious instruction, devoting as much as an hour a day to it. At the age of eight I became a devout Catholic, praying daily as I was required to do, and even serving as an altar boy during Holy Mass at the orphanage chapel.

Classes in religion were taught not by the nuns, but by the priest, who I also remember as kind and caring. There is little I remember from my Catholic religion classes, but one "fact" stayed with me—that the Jews were responsible for the crucifixion of Jesus. I also recall that our priest-teacher asked us to save silver foils from candies and press them into shiny balls that would help missionaries convert African children to Christianity, so they too could go to heaven. I obliged.

While at the orphanage, I was not completely separated from my family. During my stay there, my parents were able to move from Tehelné pole to an apartment in a nicer neighborhood on a hill known as Kalvária. The new residence was a one-bedroom apartment, not even half the size of our prewar apartments, but it was in a two-family house in a pleasant residential section, only a fifteen-minute walk from the orphanage. Our neighbors, the Blagodarnys, émigrés from Communist Russia, were kind to my parents and to me. On weekends, when I came home, our neighbors would prepare authentic Russian pierogi—small dumplings filled with ground meat—that they often shared with us.

During the time we lived in the Kalvária Hill district, the Blagodarnys went to the movies at least once a week. They were not movie buffs; in fact, they would usually leave the theater before the main feature was screened. They went to the cinema in those pre-TV days to see the weekly news journal that always preceded the feature movie. That news regularly showed the German army, then still victorious, advancing on Russian soil. The Blagodarnys had no warm feelings for the Germans (or the Soviets for that matter), but they were dying to catch glimpses of their beloved Russian homeland.

During this period, my maternal grandparents were evicted from their apartment too and had to move into my parents' Kalvária Hill apartment. With Grandpa and Grandma occupying a maid's room in the one-bedroom apartment, the space was tight. Nevertheless, I enjoyed the time I spent with my family and Kalos, when, after the end of school on Saturdays, I could walk home to Kalvária Hill. I also spent much of the summer vacation after the third grade with my parents.

There were some tense moments between my parents and me precipitated by the Catholic religious education I was receiving. My parents were not required to wear the Jewish star, but my grandparents, who apparently could not receive a legal exemption, had to display the yellow stars whenever they left the house. So as not to confuse me, my grandparents tried to hide their stars from me, but of course, they were not successful. "Are we Jewish or are we Catholic?" I demanded to know, pointing to my grandparents' overcoats.

At the orphanage school I was brainwashed not only to become a devout Catholic, but also to be a loyal citizen of the pro-Nazi Slovak State. I suspect that most of the St. Vincent Sisters were not Nazi sympathizers, otherwise they would not have protected Jewish boys, but government regulations required that educators instill in their pupils a feeling of loyalty toward the pro-Nazi regime. One day I was taken with my entire class to wish a happy birthday to Vojtech Tuka, the premier of Slovakia and one of the most outspoken supporters of the eradication of the entire Jewish population.

As my ninth birthday was approaching, a Jewish friend of our family asked me what present I would like to get for my birthday. Without hesitation I said that I would like to have a Hlinka Youth organization uniform. The Hlinka Youth, named after the Slovak nationalist Andrej Hlinka, was the Slovak equivalent of Hitler Youth. My parents and their friend turned pale, but were afraid to say anything to me. I did not get the uniform.

Another present I desperately coveted but never received was an electric toy train. My second cousin Hansi had a toy train, complete with ramps and tunnels, and I was dying to have one too. My parents never bought me a toy train. My cousin Hansi was taken to Auschwitz in 1944, where he perished.

—

Early in 1943 my mother was assigned to run an ophthalmology office in Prievidza, at the time a town with a population of five thousand in west-central Slovakia, some hundred miles away from Bratislava. After her move to Prievidza, my father continued to work in Bratislava, living with my maternal grandmother in the Kalvária Hill apartment. My grandfather had died of natural causes at the beginning of 1943 when I was still at the Catholic orphanage boarding school.

An event that occurred in the spring of 1943, along with my mother's transfer to Prievidza, led to my parents' decision to take me out of the orphanage. A reporter came to the orphanage to document the lives of the orphans and the laudable work done by the St. Vincent Sisters. Soon an article was published in a Slovak weekly magazine, including a photo of me. I was not identified by name, but my parents were worried that someone would recognize me and realize that I was being illegally protected. In fact, the need to keep me in the orphanage at that moment in time was no longer pressing because Jewish transports from Slovakia to concentration camps ended in the fall of 1942 and my parents had secured legal exemptions from the discriminatory anti-Jewish laws. My parents made the decision that I would join my mother in Prievidza. My father would continue living in Bratislava, where he was allowed to work.

I finished the fourth year of elementary school while at the orphanage, and with my good grades, I had the option to skip fifth grade and apply for acceptance to a gymnázium. There was a public gymnázium in Prievidza and they were ready to take me. I assume that despite the laws prohibiting Jewish children from attending regular public school and especially any school above elementary school level, I was accepted to the Prievidza gymnázium because of my parents' and my exemptions. I recall, too, that most people in Prievidza, who must have known of my mother's and my Jewish ancestry—after all, it was a small town in a small country—were extremely kind to us. It is remarkable that during the year-plus we spent in Prievidza I don't recall experiencing a single anti-Semitic incident.

In Prievidza my mother and I lived in a small apartment in a one-story house. The larger room was reserved for my mother's ophthalmology practice, but my mother also slept there on a sofa. The smaller room served as both a living room and my bedroom. My mother befriended two families in Prievidza. One was a Jewish general practitioner, Dr. Ján Neumann, and his non-Jewish wife, Helena, a teacher at the local gymnázium. Helena married Dr. Neumann, a widower, in 1941 or 1942, thus protecting him from being sent to a concentration camp.

The second family my mother befriended were the Heumanns, fully Jewish, and, until the Fascist takeover, the wealthiest people in Prievidza. The Heumanns had owned a successful company called Carpathia that made fruit preserves; like all Jewish-owned enterprises, the company was Aryanized, but the former owners were still actively managing the business. My mother and I were frequent guests of the Neumanns and the Heumanns in their affluent family homes. At least outwardly, these two families appeared to be living fairly normal and comfortable lives. And this was in 1943, only about a year after two-thirds of the Slovak Jewry had been deported to extermination camps. I should add that in 1941 Prievidza had a Jewish community of several hundred people. By the time my mother and I moved to Prievidza, most of them were gone, and probably no longer alive. Our modest but relatively safe life in Prievidza in the midst of the Second World War is but one example of the many paradoxes that existed in Slovakia in those days.

I completed the first grade of the gymnázium at the end of June 1944. Less than two months later our lives were once again torn apart by the surrounding circumstances.

An armed insurrection against the pro-Nazi Slovak government of Jozef Tiso was launched on August 29, 1944. The Slovak National Uprising, as the insurrection came to be known, was organized with the support of Edvard Beneš, the president of Czechoslovakia, and his exiled government in London. The nucleus of the uprising was dissident groups of the Slovak army and members of the Slovak intelligentsia opposed to the Nazis. Their actions converged around the town of Banská Bystrica in central Slovakia. At about the same time the German Wehrmacht and SS units started occupying Slovakia from the west and north. Bratislava remained in the hands of the Tiso government, but central Slovakia, including Prievidza, fell under the control of the insurgents.

My mother and I were overjoyed when suddenly Czechoslovak flags featuring a blue triangle wedged between white and red stripes were hoisted on some buildings in Prievidza. My father caught the last train out of Bratislava to Prievidza, before all transportation between government-controlled and rebel-held territories was halted.

Our joy was short-lived. In a few weeks it became clear that the rebel forces were too weak to resist the onslaught of the German army and the special Slovak units still loyal to the pro-Nazi government. My parents knew that we couldn't stay in Prievidza because the Germans and their Slovak allies would not recognize the legal exemptions issued by the Tiso government.

We consulted with our friends, the Neumanns. Dr. Neumann had already some years earlier built an underground hiding place for himself and his elderly father somewhere deep in the woods and had arranged with reliable non-Jewish friends that they would bring him food and other essential supplies. Dr. Neumann's wife, though not Jewish and therefore not in immediate danger of being killed or deported, did not want to stay in Prievidza either, lest she would be interrogated and forced to reveal her husband's hiding place. Mrs. Neumann knew a family in a remote hamlet near the village of Valaská Belá, some twenty-five miles away from Prievidza. It was decided that my mother and I would seek shelter in Valaská Belá together with Mrs. Neumann.

My father decided not to come with us, fearing that the presence of a forty-eight-year-old man would arouse suspicion, since most men that age were working or serving in the army. Instead he decided to retreat with the rebel military units to the town of Banská Bystrica, the epicenter of the uprising.

The hamlet where we sought shelter consisted of only half a dozen dwellings, occupied by peasant families. The place was truly remote, about a mile-long climb from the highway up a steep hill. It was not accessible by car and only with great difficulty by a horse-drawn carriage. The families living there supported themselves by primitive farming on tiny plots of land and by raising farm animals—cows, pigs, geese, and chickens. Houses in Slovak villages in those days had no plumbing, no telephones, and often no electricity. The house we moved into had dirt floors. My mother and Mrs. Neumann were a bit shocked but determined to make the best out of it. At age eleven, I didn't mind having no bathroom and not taking baths—it was like being at a summer camp.

It was late September and the weather was still mild. To make myself useful I offered to take the landlady's cows out to pasture. The offer was accepted. I was instructed where to take the cows and when to bring them back home to their stables. After only a few days I considered myself an expert cow-boy, until something unexpected happened.

It was not unusual in those days to see British or American warplanes flying to or from a bombing mission. As I was watching the cows grazing peacefully, I noticed a squadron of bombers flying, as usual, in the relative safety of high altitude. Then I heard the roar of a nearby explosion as one of the planes dropped several bombs over empty land. I had no idea what was going on, or whether the next bombs would be dropped right on top of me. In a terrified dash for shelter, I lost sight of my cows and when I recovered from the shock, I saw the panicked cows running away madly. I waited, hoping the cows would return to the same pasture, but they did not. I was petrified anew. How would I explain to our landlady that I had lost all her cows? Finally, not knowing what else to do, I dragged myself the one- or two-mile distance back to the hamlet. To my astonishment, all the cows were back in their stables. They had returned home without my help.

My mother, Mrs. Neumann, and I did not stay in the hamlet much longer. There were rumors of armed clashes between the retreating insurgents and forces loyal to the Tiso government in the surrounding area, and my mother felt that our shelter was not safe. We packed up and moved to Nitrianske Rudno, a village about ten miles down the road. We did not know anyone there, but we found a Jewish dentist, who—to our surprise—was still living openly in his own house. My mother asked if he could recommend a reliable person or family who might be willing to take us in and protect us. The dentist recommended Tomáš and Mária Javorček, a couple who had recently returned to the village after living and working for many years in pre–World War II France. He told us that after the German invasion Tomáš had fought the Germans in the French army and he hated the Nazis. Not being Jewish, Mrs. Neumann decided to take a chance and go back home to Prievidza, but my mother and I moved in with the Javorčeks.

Theirs was an old but unusually large house, a former manor, bought with the couple's savings upon returning from France where, before joining the French army, Tomáš was a factory worker and Mária a maid in the household of a rich Parisian family. The house was not only one of the largest in the village, it was also situated smack at the intersection of two highways—not an ideal hiding place. But my mother was relieved that we had found a sympathetic family and paid little attention to the location. My mother still had some cash and a fee for room and board was agreed upon. The Javorčeks proposed that we use their spacious bedroom; they would be sleeping with their two small children in a den. Their house too lacked indoor plumbing, but it had a handsome wooden floor, and it was quite comfortably furnished.

———

We stayed with the Javorčeks from mid-October 1944 through our liberation by the Soviet army on April 4, 1945, living more or less in the open. The official explanation of why we came to live in Nitrianske Rudno was that we left our home in Bratislava in order to escape air raids— not an unlikely story since Bratislava was an industrial center frequently bombed by Allied warplanes. My mother had an official Slovak identity card issued in her own name that listed her religion as Roman Catholic. I am not sure if this was a legitimate document to which she had been entitled because she was "exempted" from being considered Jewish, or whether this was a counterfeit identity card bought on the black market. As an eleven-year-old child I didn't need an identity card.

In order to blend in with the local population and to resemble Slovak peasant women around her, my mother started wearing a kerchief over her head. The Javorčeks certainly knew that we were Jewish. Did other people in the village know? I believe that at the very least they must have suspected it. It attests to the decency of the people of Nitrianske Rudno that no one reported our suspicious presence to the authorities. Members of the occupying German forces came through town regularly enough, and a single word to one of those soldiers could have meant the loss of our lives. Nor did anyone report the Javorčeks for illegally harboring Jews—an offense severely punishable at the time.

At the beginning of our stay in Nitrianske Rudno not much was happening. My mother was helping Mrs. Javorček with household chores. The Javorčeks owned a vegetable garden and a small field where they grew corn. They were also raising chickens, geese, and pigs. Homegrown vegetables, eggs laid by the chickens, and meat from the farm animals provided much of the daily food; fresh milk we had to buy from the neighbors who owned cows. Twice a week Mrs. Javorček would knead dough to make homemade bread, later baked in a special oven by one of the neighbors, the designated baker.

I did not attend classes because the village had only an elementary school, for which I was too old. Commuting fifteen miles to the gymnázium in Prievidza—even if it were technically feasible—was out of the question as I would be immediately recognized there. Since there were no cows for me to take to pasture, I spent my days reading whatever books and periodicals I could lay my hands on, and playing with the neighbors' boys, quite happy not going to school. I remember being bored. I missed my father and grandmother. Grandma had stayed behind in Bratislava when my father departed in a hurry to join my mother and me in Prievidza just before the German troops had moved into Bratislava. We had no news from either one of them since leaving Prievidza in September.

As the winter approached the relatively tranquil environment of our first weeks in Nitrianske Rudno began to change. Initially, it was the partisans fighting the Germans who would start paying us visits, sometimes in broad daylight, sometimes at night. Because our house was so large and easily accessible from the intersection of two highways, they often knocked on our door—automatic rifles strapped around their shoulders—eager to come in to warm up and have a conversation. The Javorčeks and my mother generally welcomed them, and we would offer them a warm drink and words of encouragement before they returned to their hideouts in the woods. But soon less welcome visitors—German army soldiers—started to show up in the village. They too found their way into the Javorček house. These were ordinary members of the Wehrmacht, army soldiers, not the fanatic members of the SS troops. Yet it is an understatement to say that my mother was uncomfortable seeing German soldiers.

At age eleven, I viewed the German soldiers with an equal measure of dread and fascination. My mother instructed me never to reveal that I could speak German—one of the three languages I knew by the age of four—in front of the soldiers. Then one day a group of three German soldiers came into the house to seek respite from the cold weather and to rest up. They sat down in the kitchen, the only warm room in the house, heated by a wood-burning kitchen stove. Tomáš and Mária, their two small children, my mother, and I were also in the kitchen, mostly silent. One of the German soldiers, I think a noncommissioned officer, finished smoking and was looking for a place to extinguish his cigarette. "*Da unten*" ("here down below"), I said, pointing to the grates at the bottom of the burning kitchen stove. Then I froze in panic. "Ah, you speak German?" the soldier inquired. "Learn German in school," my mother tried to explain, pretending to know only a few words of German. I remember the surprised expression on the German's face. The air was thick with tension. Then the soldiers departed without an incident.

There were numerous other difficult moments. One day we saw armed

German soldiers marching a family, two or three adults and a couple of kids, almost certainly Jewish, up the highway in front of our windows. We didn't know who they were or how they were found out. Later we learned that the whole family was shot dead in the neighboring village.

Another day, two partisans entered the village while several German soldiers were visiting the local pub only two hundred feet from the Javorčeks' home. Local residents warned the partisans about the presence of the German soldiers, but instead of retreating, the partisans decided to use the situation as an opportunity to inflict casualties on the Germans. We watched from the window of our house as they positioned themselves behind a fence, their Russian-made automatic rifles aimed at the entrance of the pub. As the German soldiers exited the pub, the partisans opened fire. They missed, enabling the Germans to rapidly retreat back into the pub. The two partisans, young and apparently inexperienced in matters of warfare, continued waiting in their hideout. Within minutes, German reinforcements arrived in two armored vehicles, shooting and killing the two partisans—all of this in our plain view. The dead bodies of the partisans were left lying in a ditch for the rest of the day, as people were afraid to come out and move them.

If a similar incident had happened in 1941 or 1942, rather than at the beginning of 1945, the entire population of Nitrianske Rudno would likely have been held responsible and severely punished by the Germans. By this time though, with the Russian army already advancing in eastern Slovakia, the Germans seemed more intent on saving their own skins than punishing the local population.

Around the same time Nitrianske Rudno was also visited by Hungarian troops fighting alongside the Germans. On a sunny, springlike day in February 1945, two Hungarian soldiers were resting in the Javorčeks' yard while I was visiting the outhouse. As I passed the two soldiers—who would never have suspected that I understood Hungarian—one of them, pointing at me, said to his companion, "*Nézz ide, én biztos vagyok hogy ez egy Zsidó gyerek*" ("Look here, I am sure this is a Jewish child"). After seventy years I still remember how scared I was when I heard this comment. Would my mother and I now be marched to our death like the Jewish family a few weeks earlier? I quickly returned to the house and closed the door behind me. The Hungarian soldiers departed moments later.

⸺ ◆ ⸺

After living in Nitrianske Rudno for about three months my mother ran out of money. Nor did she have any jewelry or other possessions to pawn or sell. When she told the Javorčeks that she could not continue paying for our room and board, they said that was fine, we could stay with them anyway. But my mother felt uneasy accepting the offer because the Javorčeks were unemployed (as were most other people in the village at the time), and they had no source of income. Although they produced much of the food they and we consumed on their small piece of land, they needed some cash to pay for flour and a few other staples.

My mother concocted a bold plan. She had a gentile friend in Bratislava, Mrs. Subak, an ethnic German, who my mother thought would still be sufficiently well-off to be able to help. My mother prepared a handwritten letter to her friend, explaining our situation and asking if Mrs. Subak could lend her money that would be sufficient to tide us over through the end of the war. Of course, my mother could not send the letter by mail because it could have gotten into the wrong hands, so she asked Tomáš Javorček to travel to Bratislava by train (a nearly three-hour journey each way, including a long hike to the train station), find Mrs. Subak, give her the letter in the strict privacy of her home, and if she were to agree, bring the cash back with him to us. As unlikely as it sounds, the plan worked and, as a result of Mrs. Subak's enormous generosity, Tomáš returned with a substantial enough sum of money that allowed us to pay the Javorčeks and any other expenses through our liberation by the Soviet army. After the war, my mother repaid the loan and thanked Mrs. Subak in person for her courageous act of friendship and humanity.

By February 1945, we could hear the roar of heavy artillery fire as the front line moved closer to us. Our hopes for the fall of the Nazi regime were rising by the day. At the same time my mother had told me that the final weeks could be treacherous. Visits by German soldiers to our house became more frequent. At one time my mother and I had to move out of the large bedroom and sleep in the kitchen when a group of Wehrmacht officers requisitioned the space as their quarters for a few days before their unit had moved on. We did have the feeling that they viewed us with some suspicion, because despite my mother's kerchief and my worn clothes, neither one of us looked like we belonged there. I had dark curly hair—uncommon among Slavic people—and if the Hungarian soldier recognized that I looked Jewish, so could the Germans. My mother's features, though perhaps not very Jewish, also left no doubt that she was no peasant, with or without a kerchief covering her hair. In the end they left us alone.

Finally, on April 3, the last German troops departed, heading north toward Valaská Belá. On April 4, the first reconnaissance unit of about eight or ten Russian soldiers, including some women, with automatic rifles at the ready, walked up the highway in front of our house. Tomáš Javorček, unable to suppress his joy and excitement, came out of the house, infant child on his arm, to welcome them. Despite my mother's effort to prevent me from leaving the house, I followed Tomáš. "*Na Berlin*" ("On to Berlin"), shouted one of the soldiers. "*Na Berlin*" was Tomáš Javorček's response. The six long years of waiting for the end of the Nazi rule were finally over.

Not everything was roses. After the first reconnaissance unit had moved on, more Russian troops arrived. Of course, they would not miss coming to the prominently situated Javorčeks' house, asking for bread and vodka. We obliged with the food, but not the drink, not wanting a houseful of drunken, rowdy soldiers; I don't think the Javorčeks had any vodka anyway. The soldiers looked bedraggled, tired, and hungry. They were all very young, and judging from their facial features they must have hailed mostly from the Asian regions of the Soviet Union.

Later that day a Russian officer arrived and immediately became overly friendly to my mother. He announced that he would return at night. That, my mother and the Javorčeks knew, was a dangerous prospect, and as soon as the officer left, my mother decided that she had to spend the night elsewhere. For some reason it was decided that my mother and I would hide in a nearby brick factory. The factory had a large enclosed hall, which to our surprise was filled with people, mostly families, who all looked like refugees. We did not know who these people were or for how long they were staying there, and we did not pry. We lay down on the floor, covered ourselves with the blankets we brought with us, and passed a sleepless night. Mrs. Javorček came in the morning to tell us that the Russian officer had been outraged that my mother was not there when he came back, but eventually left without harming anyone. It was now safe to return to the Javorčeks' house.

We stayed in Nitrianske Rudno a couple more days, but my mother was restless. She did not know what had happened to her mother, who had been alone in Bratislava since my father had left there in a hurry in August of the preceding year. About my father we knew only that he left with the retreating insurgents in the general direction of Banská Bystrica, but we had no idea where he was or whether he was still alive.

Even though getting on the road was not entirely safe—there was a great deal of army traffic and Russian soldiers could be unpredictable— my mother decided to return to Prievidza where she would try to find friends. There was of course no public transportation in those days, but my mother packed our belongings into two small bags and found a friendly (though not overly friendly) Russian soldier who offered to take us to the small town of Nováky, about halfway between Nitrianske Rudno and Prievidza; from there we would have to find some other mode of transportation to reach Prievidza.

We said an emotional good-bye to the Javorčeks, climbed onto the open Russian army truck, and departed. There was heavy military traffic, and the nine-mile journey to Nováky took almost an hour. We got off at a three-way highway junction. My mother and I were standing by the highway contemplating what to do next, when, only a few minutes later, a civilian passenger car stopped, seemingly for us. The door opened and out stepped my father.

———

Lloyd Old, a prominent immunologist and cancer researcher. He and his colleagues were the first to describe tumor necrosis factor (TNF) in 1975. Old's hope that TNF would become useful as a therapeutic drug for cancer was shattered when it became clear that the substance is too toxic to be administered to humans.

Hubert Schoemaker, cofounder of Centocor, Inc. and its long-time president and CEO. My friendship with Wall and Schoemaker was key in the collaboration between my laboratory at NYU School of Medicine and Centocor that resulted in the development of the therapeutic drug Remicade. (Photo supplied by Anne Faulkner Schoemaker.)

Michael Wall, cofounder and long-time chairman of Centocor, Inc.

My mother
Friderika Fischer
c. 1930.

My father Július Vilcek c. 1930.

My poodle Kalos (age 2) being pampered by me (age 1).

With my mother c. 1935.

First communion at the Catholic orphanage where I was hidden c. 1942. I am in the center of the last row. My Jewish friend Jan Deutsch is in the last row on the far left (near the priest).

Former manor house in the village of Nitrianske Rudno where my mother and I lived during the final months of the Second World War.

With my mother c. 1946.

Robert (Bobby) Lamberg
c. 1948.

Alex (Šaci) Vietor (in the driver's seat of his motorcycle) and I c. 1958. Bobby
and Šaci were my closest friends during the postwar years in Czechoslovakia.

I am examining bacterial cultures as a budding scientist at the Department of Microbiology and Immunology during my medical studies in Bratislava c. 1955.

At the Institute of Virology c. 1958. Seated: Helena Łibíková, department head, and Dionýz Blaškovič, director. Second row l. to r.: Jan Závada, Josef Řeháček, and me.

Marica Gerháth, my wife-to-be, in the Tatra mountains c. 1962.

Alick Isaacs, British scientist and co-discoverer of interferon. Isaacs's lecture I had attended in Bratislava spawned my interest in the study of interferon.

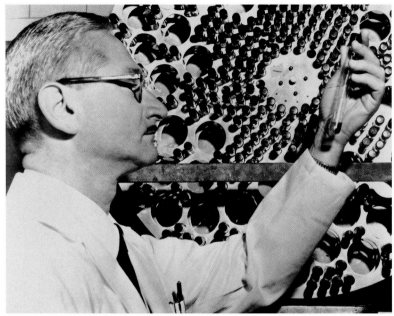

Albert Sabin, American scientist best known for the development of the live polio vaccine. Advice Sabin gave me during a visit to Bratislava was instrumental in my decision to investigate interferon.

I am chatting with Jaqueline and Edward DeMaeyer (right) at the first international conference on interferon I had organized in Smolenice near Bratislava in 1964.

My Škoda-Spartak automobile c. 1962. This was the car in which Marica and I defected from Czechoslovakia in the fall of 1964.

We learned that after the uprising had been suppressed in October 1944, my father walked through woods and small country roads toward the Russian front. Once he was stopped by a German patrol and taken in for questioning, but somehow he managed to extricate himself. He continued walking eastward and by mid-December succeeded in crossing the front line to the Russian army and reached the liberated town of Užhorod in the region of Subcarpathian Rus that until 1939 was part of Czechoslovakia but is now in Ukraine. There my father offered his services to the Czechoslovak government in exile.

By January 1945, the temporary government moved to the liberated eastern Slovakian city of Košice, and my father moved with them. When he learned that Prievidza had been secured by the Russians in early April, he found a car and a driver (not an easy feat in those days) to take him the 170-mile distance over war-torn roads from Košice to Prievidza, determined to try to locate us in the midst of the postliberation turmoil. He, too, had no idea whether we were alive and how we had spent the seven months since our separation. Someone in Prievidza told him that we might be in Nitrianske Rudno. He was heading to Nitrianske Rudno when he spotted us on the highway. The joyous reunion with my father on an open highway, amidst chaotic circumstances, after a long winter of separation, stands out as one of the most vivid memories of my life.

There was no such happy reunion with my grandmother. As we learned only much later, my grandmother was rounded up in Bratislava in October 1944 and taken to a concentration camp. My mother was never able to find out where or how she died. Some survivors reported that she was taken to Ravensbrück, a notorious women's concentration camp in northern Germany, where, given the history of that camp, she very likely died of typhoid fever in early 1945.

Chapter Six

Peacetime Under a New Regime

We did not get back to Bratislava until the end of June 1945. From mid-April until our return to Bratislava we lived in the eastern Slovakian town of Košice, where my father worked for the provisional Czechoslovak government and I—after a hiatus of nearly one year—attended school.

People were still trying to figure out who had and who had not survived. The sight of former concentration camp inmates tracing their way back to their homes was common. There were shortages of everything. But we lived once again in a free democratic Czechoslovak Republic, and among the survivors there was initially widespread optimism that things would improve.

Once back in Bratislava my parents began to piece their life together. We were able to rent a spacious apartment in an early-twentieth-century three-family home in a pleasant villa district, not far from the center of the city. We even recovered some of the furniture and Oriental carpets my parents had left with non-Jewish friends and acquaintances for safe-keeping.

Kalos, our poodle, had survived the war. My father had asked our former cleaning lady to care for Kalos, which she gladly agreed to do. My mother and I came to claim him after returning to Bratislava in the summer of 1945, but the woman started crying, saying that she could not imagine life without Kalos. By then he had reached the ripe age of thirteen and, sadly, he seemed indifferent to my mother and me. We did not put up a fight, deciding it was probably best to let Kalos live out his life with his new mistress.

My mother once again hung out a shingle, using one room in the apartment as her ophthalmology office. My father recovered his position in the company that emerged after the war as the successor of Hand-lovské Uhoľné Bane, the Handlová Coal Mine Company. In the fall of 1945, I was admitted to the third grade of a gymnázium, even though I had missed most of the second grade. I became a boy scout and made some friendships that have lasted to this day.

One of my lifetime friends whom I met as a boy scout is Robert "Bobby" Lamberg, a Holocaust survivor from a German-speaking Jewish family. The friendship with Bobby was unusual because he was four years older than me. When I told my mother about my new friend, she was worried that having a seventeen-year-old friend at age thirteen ran the risk that he would drag me into activities for which I was not ready. However, when I introduced Bobby to my mother she reconsidered; later she told me she had realized that if anyone was going to be a bad influence, it would be me on Bobby.

Bobby had had a difficult life. His parents were well-to-do when they lived in their home in the Sudetenland region of Czechoslovakia. The Lambergs fled their home before the annexation of Sudetenland by Germany sanctioned by the Munich Agreement, and eventually ended up in Bratislava. Bobby and his parents survived the war by not registering as Jews, but Bobby's two older brothers were caught and they perished in concentration camps. Bobby's father had a doctorate in philosophy from the University of Vienna and he was a veritable Renaissance man. He spoke at least half a dozen languages, though Slovak or Czech were not among them. By the time I befriended Bobby, the family was quite impoverished, living in two rooms of a communal apartment they shared with two other families. In his seventies by then, the only income Mr. Lamberg had came from giving private lessons in French and English. German was Mr. Lamberg's first language, but no one in those imme-diate postwar years was interested in learning German.

One of the benefits of my friendship with Bobby was that I relearned German. Although it was the first language I acquired as a child, I had heard little German and had had no opportunity to speak German since age five. As Bobby's parents spoke no Slovak or Czech, I had to com-municate with them in German. To my surprise, my German returned quickly. Another benefit of our friendship was that through Bobby's family I became exposed to an intellectual environment I had not experienced before. Though sparsely and modestly furnished, the two rooms inhabited by the Lambergs overflowed with books, all neatly covered in protective off-white paper wrappers, affixed with adhesive labels featuring the handwritten name of the author and book title.

Mr. Lamberg spent most of the day reading, actually studying, the books in his library, which ranged from classical German poetry to tracts on history and philosophy. My mother and I too liked reading books, but for Mr. Lamberg, books were the essence of his life. And the books Mr. Lamberg was reading—Goethe, Leibnitz, and Kant, among others—were quite different from the more popular type of literature I was used to at home.

Through my friendship with Bobby I also became exposed to classical music in a way I had not experienced in my own home. One of Mr. Lamberg's favorite activities, besides reading, was humming or whistling classical tunes of his beloved composers. He was partial to Mozart and Schumann; later nineteenth-century composers, such as Brahms or—god forbid—Wagner, he liked less. My parents enjoyed listening to music, especially of the more popular genre, but their musical tastes in classical music were not particularly discriminating and neither one of them could carry a tune. My own musical talents are not better. I like to listen to music—classical, opera, and jazz—but my ability to discern subtleties of music is limited and I can't abide karaoke bars.

Bobby's parents passed away in the early 1950s and in 1957 Bobby defected. After his defection, Bobby spent several years wandering the globe, eventually becoming a successful correspondent for the most prominent Swiss newspaper, *Neue Zürcher Zeitung*. After more wandering around the globe as a correspondent, he is now living in retirement in Buenos Aires with his Argentinean wife. We still speak on the phone regularly and see one another when the opportunity arises.

In addition to relearning German, courtesy of the Lambergs, I was also taking lessons in English. My mother at first arranged for me to take private lessons with a native Englishwoman who had married a Slovak man.

Apparently, I was lax in doing my homework and not making enough progress. After a few months my English teacher called my mother. "You are wasting your money," she said, "your son will never learn English." Undeterred, my mother found me another English teacher.

<hr />

After the war, Edvard Beneš, the exiled president, once again became president of Czechoslovakia. All Soviet troops departed from Czechoslovakia by early 1946, raising hopes that the country would be able to stay its course toward a free democratic future. Immediately after the end of the war there was a tilt toward a more restrained form of capitalism. Some large companies—including the coal mine company my father worked for—were nationalized and farms owned by large landowners were parceled out among smaller farmers. In general though, most forms of private ownership and private businesses were initially left intact. In 1946, the first free elections were held after almost ten years.

Though they did not win a majority, the Communist Party emerged as the strongest political party in the country—though not in Slovakia—and they ended up dominating the central government. The head of the Communist Party, Klement Gottwald, a faithful disciple of Stalin, was named prime minister of Czechoslovakia. The Communists wasted no time in manipulating the system to place their own people in positions of power. Over the years that followed their manipulations escalated, until they took over the government completely and imposed yet another totalitarian system in February 1948. The circumstances surrounding these events are extensively documented in many publications, including the illuminating memoir *Prague Winter* written by the former US secretary of state Madeleine Albright, who was born in Czechoslovakia.

By the time the war ended I was old enough to formulate my own political views. My political leanings were initially unsettled, as were those of my parents. We had some sympathy for the Communists, because during the war they strongly opposed the Nazis. I was also attracted to the idea of social justice that was, at least on paper, at the core of the Communist ideology. When it became clear that the Communist Party was more interested in usurping power than in promoting social justice, my views changed. I kept a diary for a while, and in it I remember expressing my strong opposition to the February government takeover, dubbed "victorious February" by the Communists and "Communist putsch" by the opposition. But then, worried that my diary might be discovered by the authorities, I burned it in our coal stove.

Like the rise of Nazism in the 1930s, the Communist takeover precipitated a wave of emigration. (Charles Simic, the Serbian-born American poet, famously observed, "Hitler and Stalin were my travel agents, if they weren't around, I probably would have stayed on the same street where I was born.") Some of our acquaintances who were politically active in parties opposed to the Communists decided to leave during the first weeks after the takeover, while the borders were still open. Soon, however, the government clamped down on international travel, forcing people who were desperate enough to leave to cross the border to West Germany or Austria illegally—a feat that would become increasingly more risky as border fortifications along the emerging Iron Curtain were erected.

As an accident of history, for the few Jewish people in Czechoslovakia who survived the Holocaust the window of opportunity to legally emigrate stayed open a little longer. The Soviet Union strongly supported the establishment of the State of Israel, hoping to thus counter the British influence in the Middle East. When Israel declared independence in May 1948, the Soviet Union was the first country to officially recognize the new Jewish state, and Czechoslovakia became one of the major suppliers of arms to Israel. In view of the good relations, the Czech Communist government initially allowed Jews to emigrate to Israel. Two of my father's cousins who had survived the Holocaust decided to emigrate. I was quite close to them then, because during the summer of 1947 I had spent several weeks at their home in the south Slovakian town of Levice. My mother's best friend Lilly Wertheimer also decided to leave for Israel, with her gynecologist husband and teenage daughter.

For a while my parents considered emigration to Israel, but, like in 1938, they could not make up their mind. At one point in the summer of

1948, we all sat down and started to prepare a list of belongings we would take with us to Israel, but not even that effort progressed very far. From what I can remember, I too could not make up my mind whether I would rather stay or leave. The easiest solution was to do nothing.

What was it like to live in Communist Czechoslovakia after 1948?

People active in political parties opposed to the Communist takeover were subjected to persecution. Once again there were political prisoners, people who were sentenced to harsh prison terms for being on the wrong side of the political spectrum. If you were on the wrong side—or even if you were not—you might be denounced by your neighbors. (The worst was yet to come in the 1950s when political show trials inspired by Stalinist purges in the Soviet Union would get underway.) People who owned businesses, large properties, or larger farms were also likely to experience the "iron fist of the working class." Because the country had yet to recover from the destruction of the war years, and, because the Communists grossly mismanaged their nationalized economy, there were shortages of almost everything, from food to clothes to toilet paper. In sum, life was pretty miserable for the majority of people, excepting, of course, the Communist bosses who were awarded special privileges. (And perhaps a few black marketers, provided they managed to avoid being caught.)

My family was fortunate to avoid the worst excesses of the Communist system. Unlike many other members of the upper-middle class, my parents were not forced to move into the countryside and take up manual labor. It helped that we did not own a business or even a family home, which someone in the Communist Party hierarchy might covet, so we were less of a target.

As an ophthalmologist, my mother was allowed to practice her profession, although she could no longer continue her private practice and instead had to move to a state-run ophthalmology clinic. Her workload (though not her compensation) increased dramatically, but my mother liked her work and felt appreciated by her patients. My father continued to work for the now-nationalized coal mine company, and, even though he no longer commuted to work by chauffeur-driven car as he used to before the war, he too felt appreciated.

The full-time maid we hired after moving back to Bratislava in 1945 had to leave because maids were considered a "bourgeois anachronism," and because our family's financial situation deteriorated so that we could not afford a maid in the first place. But my parents were still relishing the fact that they had survived the war. It helped that all along they had a relatively relaxed attitude toward money and material possessions—something I believe I inherited from them. When they realized they would not recover much of their prewar wealth or when they lost money during numerous "monetary reforms" carried out by the Communist government, my mother would repeat one of her favorite expressions: "It's only money."

When, in September 1945, I started attending the third grade of gymnázium in Bratislava (equivalent of eighth grade in the US), people were still preoccupied with rebuilding their war-torn lives. To obtain food and other basic goods required standing in lines, often for hours. In this atmosphere many people did not view education as the highest priority.

The quality of education also suffered as a result of changes that were being introduced to the curriculum. For example, German language instruction was replaced with Russian, but there were very few qualified teachers of Russian. Our teachers were literally just one or two lessons ahead of their students.

Within two years after the end of the war, life in general and the standard of our education had started to improve, but in February 1948, the Communist takeover once again threw life and the system of education into disarray.

In many of our classes, there was an emphasis on memorization. Remembering the years of a monarch's reign or of historic battles was more important than learning what influence a king or queen had on history or why certain battles mattered more than others. It is not surprising that we were not led to independent thinking; encouraging students to form critical opinions and ask probing questions could have gotten both our teachers and us into trouble, especially after the Communist take-over.

Even though I have thankfully forgotten much of the memorized material, I still remember some useless information I learned at the gymnázium. Unlike my mother, I took no classical Greek language classes because they were no longer offered, but I did study Latin for four years—which was useful because the Latin vocabulary and grammar helped with learning other languages. My studies of Latin have also enabled me to understand simple written and spoken Italian, even though I never studied Italian. But what good does it do that to this day I can recite Latin prepositions used with the accusative case—all twenty-one of them?

I was a good student, but I never strived to achieve the best grades in the class because I didn't want to be perceived as a bookworm. I don't remember any subjects I really disliked, but in general I enjoyed less the science subjects—math, chemistry, physics—and preferred languages, literature, history, and geography.

When the putsch marking the full Communist takeover occurred in early 1948 I was a teenager attending the fourth grade of the First State gymnázium in Bratislava, corresponding to ninth grade in the US. Changes at my school occurred fairly gradually and some were more symbolic than substantial.

In June 1948, the democratically elected president, Edvard Beneš, resigned and Klement Gottwald, the Communist leader, succeeded him. I, with a couple of my classmates, was tasked with removing the portrait of Beneš hanging on the wall of our classroom and replacing it with one of Gottwald. When we opened the frame we found that hidden under the portrait of Beneš were pictures of other former presidents, including that of Tomáš Masaryk, the founding president of Czechoslovakia, and of Jozef Tiso, the cleric who was president of the Fascist Slovak State, later executed as a war criminal. We followed the apparent tradition, and, with an impish delight, we inserted Gottwald's picture into the frame but left all the former dignitaries tucked under it.

Most of us in the class were united in our opposition to the new regime. Even though we could not express our political convictions openly, there were subtle ways in which we let everyone know where we stood. Until the Communist takeover, our teachers would be addressed by the pupils as *pán profesor* or *pani profesorka*, the equivalent of "sir" and "madam." After the takeover we were told that these titles were bourgeois anachronisms, and from now on all teachers are to be called "comrade professor." In a display of passive resistance against the regime, 90 percent of us continued to address our teachers by their old titles. Only once, when I was already at the university, did a teacher admonish me, "Please address me as comrade professor." He was a former Nazi collaborator who may have been worried that I was testing his loyalty to the Communist regime.

People, especially young people, found other ways in which to express their distaste for the Communist system. One was music. The musical form preferred by the authorities was Czech, Slovak, and Russian folk songs along with some contemporary popular music with overt ideological messages. Jazz was not banned—after all it originated in the music of oppressed African slaves—but it was barely tolerated by the regime. We adored jazz and listened to it or played it on our phonographs as often as possible.

Another way of expressing unhappiness with the regime was through clothes. Communist, especially Soviet, dignitaries were not known for wearing sharp clothes. Despite the unavailability of quality ready-made clothes, the young used creative ways to imitate the latest Western fashions. In the 1950s it was extremely narrow, short pants and striped socks for boys, tight sweaters with skirts and nylon stockings for girls. My own efforts at expressing rebelliousness through fashion were rather subdued, but a few boys at our gymnázium became really good at it. The authorities took note of the trend, and official media wasted no time in criticizing these expressions of "Western decadence."

There were many forms of political indoctrination. We were required to enter the ranks of the organization ČSM, an abbreviation for *Československý sväz mládeže* ("Czechoslovak youth organization"). One day all of us gymnázium students were sent to distribute applications for membership in the Society of Czechoslovak-Soviet Friendship; a written report mentions that sixteen streets were covered by us and that all students and professors had also signed up. There were "work brigades," days when instead of attending school we were sent to do manual labor, such as helping with the harvest of potatoes or planting new trees. There were compulsory marches and celebrations, especially on May 1, the International Holiday of Workers, or when Communist dignitaries from the Soviet Union or other "fraternal Socialist countries" were visiting town.

In the annual report for the 1949–50 school year, there were over fifty events of "civic and political education" listed that we had to attend, including a visit to an exhibition, *Stalin—Our Teacher* (attendance compulsory for all pupils); a lecture, "How Did Stalin Improve the Life of Women?" (attendance compulsory for all female teachers and pupils); or another lecture entitled "The Soviet Union's Fight for World Peace," followed by the screening of the film *In the Glow of the New Day*.

With the exception of about three kids—all of them sons or daughters of Communist dignitaries—no one in my class converted to Communist ideology, at least not before our graduation in 1951. Most of our teachers also did not believe the Communist propaganda. We could tell because when communicating to us some piece of official propaganda, they—unlike the believing Communists—would do so without any trace of enthusiasm. Yet the teachers had to be even more circumspect than us in hiding their true beliefs because any overt display of political opposition would result in a dismissal. Only one teacher and the gymnázium director were outspoken, convinced Communists, but even our hard-line director was rumored to have trouble with her daughter, whose tastes in music and clothing were of the "decadent Western" type.

Not everything in my gymnázium years was oppressive. During winters the entire student body was taken on a ski holiday to the mountains in central Slovakia. I enjoyed the ski trips and became a reasonably good downhill skier. I also recall a detail from one of these trips. We traveled by train at night; in the darkened passenger compartment, I used the time to make out with the only Communist girl in our class.

By age fifteen I experienced my first, entirely platonic love affair (no stolen kisses on a train this time). Eva was fifteen, the same as me, and much more mature. We did go to see a few movies together but, to my dismay, she soon grew tired of my youthful affection.

I also became friends with my classmate Alex "Šaci" Vietor, who was two years older than me, because I gained a year skipping fifth grade, whereas he lost a year when he immigrated to Czechoslovakia from Hungary and had to learn Slovak in order to pass admission requirements. Šaci had spent his childhood years in Budapest. He lost his father in the Holocaust, and his mother subsequently married an attorney from Slovakia who was hiding in Budapest during the war. After the war they all moved to Bratislava. Like Bobby, whom I befriended a few years earlier, Šaci became a lifelong intimate friend. ("Šaci" is a phonetic Slovak spelling of the German nickname "Schatzi," derived from the word *Schatz*, meaning "treasure.")

Šaci and I would share the same twin school bench for three years. From time to time our Slovak language teacher would give us writing assignments that we were required to complete by the end of the fifty-minute-long class. Šaci was born in Budapest, his mother tongue was Hungarian, and his ability to express himself in Slovak, especially in writing, was at the time still lacking. I knew that in order to cement our friendship I had to help Šaci complete his essay before I could even begin thinking of writing mine.

Just as the Communists were unable to convert us to their ideology, they also failed to do away with teenage parties. Šaci had friends his own age, one grade ahead of us, who were old enough to organize parties with girls. The parties were referred to as *čurbes* (the original Czech word means "a mess"). Here is how it worked: When parents of one of Šaci's friends were out of town, a party would be organized in that friend's home, to which an equal number of boys and girls would be invited. Šaci and his friends were popular boys, so the girls who participated were also attractive and popular. At the party, there would be some conversation, music, and dancing. Alcohol was served too—usually some wine or sweet liquor—and soon the lights were dimmed, with boys and girls, now paired off, finding a quiet place in some corner. I don't think that the intimacies ever progressed that far, but it was a lot of fun. I yearned to be invited to the *čurbesses*, and at times I was. At other times, Šaci's friends protested that I was too young and, to my dismay, I would not be included. To this day—and I am over eighty as I am writing these lines—as I recall those times I reexperience the disappointment of being excluded.

PART THREE | A Scientific
Education

Chapter Seven

Microbe Hunting

By the time I reached the eighth and final grade of gymnázium I needed to make a decision about my university studies. The European system is different from the American. High school graduates are usually a year older in Europe, but there is no equivalent of the liberal arts college and a decision about what specific field of studies to pursue generally has to be made by age eighteen or nineteen.

About a year before graduation from the gymnázium, my school offered all students a professional counseling service to help with career choices. Included was an IQ test and some other tests that were supposed to determine whether we were more suited for the humanities or for a science-related field. After my testing was done I met with a professional counselor. He told me that I had the second-highest IQ in my class. He also said that my tests did not indicate a clear propensity for a humanities- or science-oriented career. I remember him saying that he thought I would do well in almost any field. That may have been reassuring, but not very helpful for my career choice.

Career options in the Socialist system were limited. First of all, I could not be sure that with my "bourgeois" background I would be admitted to the university because applicants from working-class families were given preference. I always had a predilection for writing and for many years I had been thinking of becoming a journalist. I wrote my first poem at age eleven. At age fourteen, I had a small essay published in a Bratislava daily newspaper—for which they even paid me an honorarium—and my passion for creative writing continued throughout my gymnázium studies. However, I knew that journalism in Communist Czechoslovakia was a profession reserved to party loyalists and not a good fit for me. Law and economics were also highly politicized fields and therefore not right for me. Even though languages and literature were among my favorite subjects, I never considered becoming a linguist or a philologist.

Then there was medicine. My parents wanted me to become a doctor all along. For a long time I resisted the idea of following in my mother's footsteps and choosing a career that would please my parents. But finally, having ruled out other options, just one day before the deadline I completed my application to the Comenius University School of Medicine in Bratislava. To my surprise, I was accepted. My parents were delighted.

Medical school in most European countries takes six years, and Communist Czechoslovakia was no exception. The curriculum in my first two years of studies included courses that in America are taught at the undergraduate level, such as chemistry, biology, and physics, in addition to the usual medical school basic science courses of anatomy, biochemistry, microbiology, and physiology.

Even though most aspects of our studies could not be greatly influenced by politics, some elements did bear the stamp of Communist ideology. One was the field of genetics. A Soviet agronomist and influential Communist ideologist, Trofim Lysenko, trumpeted the view that theoretical genetics based on the widely accepted principles of Mendelian inheritance was wrong, and that the inheritance of acquired traits was much more important than hereditary transfer of stable genetic properties. Today his views about genetics have been widely discredited.

Stalin listened to Lysenko because he believed the theories advocated by Lysenko, unsophisticated as they were, would help increase crop yields and alleviate widespread famine in the Soviet Union in the 1930s. What had endeared Lysenko to Stalin was that early in his career Lysenko was apparently successful in preventing the loss of wheat seedlings to the frosty Russian springs by first treating wheat seeds with cold and moisture. Lysenko then went on to claim that the increased resistance of the wheat would be transferred to the progeny of the cold- and moisture-treated wheat seedlings—a claim doubted by most experts.

Stalin's support of Lysenko's theories was responsible for the fact that "Lysenkoism" started to be extended to human genetics and was accepted as the official theory applicable to much of biology in all countries of the Communist bloc. When I entered medical school, the terms "gene" and "genetics" were practically banned, except when referred to in official propaganda as "bourgeois pseudoscience of Mendelism-Morganism." The main individuals targeted by Lysenkoists were Gregor Mendel, a nineteenth-century Augustinian friar who worked in the Moravian capital of Brno, considered the founder of the science of genetics, and Thomas Hunt Morgan, an influential American geneticist active in the first half of the twentieth century.

Another Soviet pseudoscientist whose name we heard cited much too often was Olga Lepeshinskaya, a protégé of both Lenin and Stalin. Lepeshinskaya was trained as a *feldsher* (a Russian term for a medical professional roughly equivalent to a nurse-practitioner) before the Bolshevik Revolution. For her participation in the October Revolution she was elevated to the post of a professor at the Medical University in Moscow, finally becoming a leading member of the Institute of Experimental Biology of the USSR Academy of Medical Sciences. She received the Stalin Prize and many other honors for purportedly showing that live cells could be generated from subcellular components, for example from egg yolk.

The evidence Lepeshinskaya had presented was later shown to be fraudulent, but, in the meantime, we were taught that Virchow's (the famous nineteenth-century German scientist's) concept that only a living cell can produce another living cell was a bourgeois lie to be replaced with the correct dialectical-materialistic-Marxist-Leninist-Stalinist theory on the origin of all living cells from nonliving matter. Lepeshinskaya's final scientific contribution was to contend that baking soda baths promoted rejuvenation, resulting in a shortage of baking soda in Eastern Bloc countries. The soda baths may have worked for Lepeshinskaya; she died in 1963 at age ninety-two.

A postscript is in order here. Lepeshinskaya's experiments were primitive and fraudulent, making her a laughing stock for me and my fellow medical students as early as the 1950s, but the idea that it might be possible to artificially create a living cell is one day likely to be proven correct. One important step in the quest to create synthetic life was taken in

2010 by J. Craig Venter—the scientist who headed a team that sequenced the entire human genome—when he synthesized a large piece of bacterial DNA, inserted it into a bacterial cell, and showed that the synthetic DNA could act as a substitute for the cell's own DNA, thus creating a cell "whose parent was a computer."

When I entered medical school in 1951, a general atmosphere of fear and suspicion permeated life at the university. Teachers and students suspected of insufficient devotion to the Communist regime were investigated by specially appointed lustration committees. These committees included students loyal to the Communist Party, whose judgment about their professors often would be influenced not only by the professor's political views, but also by the students' personal experiences. It could be dangerous for a professor to fail a student lest that student might turn his wrath against the professor, accusing him of being opposed to working-class students—a verdict that almost certainly would lead to dismissal.

One professor who refused to be intimidated by the Communist activists was our anatomy teacher, Eugénia Štekláčová, who remained strict and uncompromising in her academic standards. In those days in Czechoslovakia, anatomy was taught a full four semesters, compared to two semesters in most medical schools in the US then. Today the teaching of anatomy in the US is generally completed in less than one semester. We had to memorize the name of every single little bone in the body, every aperture in the skull, every nerve, in fact every structure in the body contained in the most detailed anatomical atlases. We also had to spend many hours dissecting cadavers under the watchful eye of Dr. Štekláčová.

At the start of my medical studies, nearly five hundred students were enrolled in my class, a truly staggering number. By the end of the second year there were about 250 students left. The majority of the students who did not make it into the third year had to drop out because they had failed anatomy. It is not surprising that Dr. Štekláčová was fired from the university shortly after my completion of the anatomy course, a victim of her refusal to compromise her academic standards along with the fact that she refused to kowtow to the Communist Party members.

Another memorable teacher was František Klein, chairman of the Department of Pathology. Klein was a Holocaust survivor. As students we heard this widely circulated anecdote about him: During the Second World War he was stopped in the street in Bratislava by members of the Slovak Fascist militia. "What do you have in your briefcase?" they asked. "Shit," Klein answered. The militiamen became enraged. "You dirty Jew, we'll teach you a lesson. Open the briefcase!" When Klein complied, the militiamen found vials filled with specimens of—feces. I don't guarantee the authenticity of the anecdote, but as the Italian saying goes, *Se non è vero è ben trovato* (Even if it isn't true, it makes for a good story).

Professor Klein was a popular teacher, and during his lectures the auditorium always filled to capacity. He had a high-pitched voice, which he raised further if he wanted to make an important point. So revered was Professor Klein that—despite his lack of socialistic ideological vigor—the Communists were unable to fire him. He stayed on until his regular retirement. Here is a tidbit from one of Klein's lectures I still remember: "When you see a patient with a one-sided inflammation of the knee, always think gonorrhea!" In addition to legitimate medical school courses, we also were required to take courses on Marxism-Leninism (meaning primitive Communist political propaganda), usually one to two hours of lectures or seminars per week throughout the six years of medical school. Attendance at these classes was compulsory, and we had to take exams showing our understanding and (faked) appreciation of Communist ideology.

Another distraction was defense education, taking up one full day per week and obligatory for all male students. In those days there was compulsory military service in Czechoslovakia; by taking the course in defense education and by attending military training camps during the summer, medical students were able to forego most of the two-year military service that graduates of other universities had to complete. (This was similar to the Reserve Officers' Training Corps [ROTC] program in the US.) With all these distractions it is a miracle we managed to learn anything at all about medical science and medicine.

During my first year of medical studies I was approached by a fellow student with whom I was assigned to work together on the dissection of a cadaver and who, I knew, was a card-carrying member of the Communist Party. Out of the blue she asked me whether I would like to join the Communist Party, adding that, based on the interactions she had with me, she thought I deserved the privilege of belonging to the party. I knew that I did not want to join the Communist Party, but was afraid to decline. I explained how surprised I was to have this generous offer extended to me. "Could I think about it?" I inquired. "Yes, of course," my colleague replied. That bought me some time, but a few days later I had to come up with an answer. "I thought about it," I said. "This is a very great honor, but I feel I am not yet completely worthy of Communist Party membership. I need to grow more ideologically and deepen my understanding of the theory of Marxism-Leninism." I was left alone and never again asked to join the Communist Party.

My closest friend during the first two years of medical school was Dušan Šťastný. A native of Prague, Dušan came to Bratislava because he was not accepted by the Charles University Medical School in Prague. The reason he was not admitted had nothing to do with academic performance—Dušan was an excellent student—the problem was his bourgeois background and that his father had served in a political post prior to the Communist takeover. Dušan was luckier with his application to the Bratislava Medical School, perhaps because his family background was not well known outside his hometown. (Dušan told me that he thinks he was accepted because he had excelled in basketball and he was expected to play for the university team, which he then decided not to do.)

Because our medical school class was so huge, Dušan and I did not attend many lectures (except Marxism-Leninism, where attendance was required). Instead, when exam time approached, we would sit down and learn the material from textbooks and transcripts. Dušan had a formidable, photographic memory. When I asked, "Where did you find this information?" he would refer me to the page number in the textbook, recalling the exact position of the sentence or the paragraph in question. I truly admired how smart he was and, yes, I was a little envious of his superb memory.

Our close friendship ended after two years, when Dušan was able to transfer into the third year of medical school back in Prague. We did keep in touch until my emigration to America in 1964. Then we lost contact with one another for thirty-five years, until a colleague of mine tipped me off that Dušan worked as a psychiatrist for the public health authority in Leicester, England.

Our reunion after so many years was joyful. It turned out that Dušan got out of Communist Czechoslovakia by volunteering to take up a medical position in Zambia, years later moving to the United Kingdom. Now living in retirement in New Zealand with his wife, Dušan has certainly had an interesting life and built a good career. Yet I can't help but think that given his terrific intelligence and talent that I had admired so much during our joint preparations for medical school exams, Dušan's lifetime professional accomplishments would probably have been more exceptional under more settled circumstances.

During my first two years in medical school there was some time for fun. My classmates and I spent a great deal of time at the coffeehouse Metropol, conveniently located in the immediate vicinity of the medical school and university hospital, popular with both faculty and students. Affectionately referred to as Metropolka, one could sit in the coffeehouse for hours and order just one coffee. We took ample advantage of this liberal policy, often spending time sitting and talking at the Metropolka instead of going to lectures. One of our favorite pastimes was playing chess; another was playing cards, which was not permitted at the coffeehouse, but we did it anyway and the underworked but poorly paid waiters generally looked the other way.

When I started my studies of medicine, I had no idea what field I would end up choosing. Sometime after the beginning of year two of the six-year curriculum I saw a posted notice that listed research projects open to medical students. One of the topics mentioned concerned the mechanism of penicillin resistance in *Staphylococcus aureus* bacteria, a project sponsored by the Department of Microbiology and Immunology, then headed by Rudolf Sónak. (Later, Professor Sónak was fired from his position at the university because of his political views. In the late 1960s he defected and became a professor at the university in Münster, Germany.) The project sounded interesting—the word penicillin still had an aura of magic in those days. Besides, like most young people of my generation, I had read Paul de Kruif's *Microbe Hunters* and Sinclair Lewis's *Arrowsmith*, and the idea of giving medical research a try appealed to me. I applied.

By the end of my second year in medical school I was hooked. I knew right then and there that what I really wanted to do in life was medical research. I enjoyed all aspects of the process, but what I liked best was immersing myself in every single publication in what I considered "my field." In those days there were only about a dozen journals important to a budding microbiologist, and I made sure to examine them all when they arrived at the Bratislava medical library, housed in an eighteenth-century building that had once served as a summer palace of Empress Maria Theresa. There were weekly research seminars sponsored by the local medical society that I attended regularly. I met famous visiting scientists from abroad, among them Hans Selye, a Canadian endocrinologist of Hungarian extraction known for his studies on the origin of biological stress, the organism's complex response to an environmental challenge.

As a third-year student I was chosen to present the results of my research on penicillin resistance of *Staphylococcus aureus* at a national conference of microbiologists. During my final two years in medical school I became interested in immunology, and I spent the summer after the fifth year of my studies working in Prague at the laboratory of Jaroslav Šterzl, a prominent Czech researcher in those days. With Šterzl's encouragement, I attended an international immunology conference in Czechoslovakia, where I listened to lectures by Australian scientist Macfarlane Burnet, the future Nobel Prize winner, and other giants in the field of immunology. These experiences reinforced my conviction that this was the world I wanted to belong to for the rest of my life. I would do everything to make this dream become reality.

———

When I graduated from medical school I had a choice: I could become a medical intern at a small provincial hospital in central Slovakia or join the Institute of Virology, a research center of the Czechoslovak Academy of Sciences in Bratislava, dedicated to the investigation of viruses. For me it was no-brainer. Virology was part of the science of microbiology that I had fallen in love with as a medical student. Like bacteria, fungi, and protozoa, viruses are disease-causing infectious agents. In fact, viruses are responsible for approximately one-half of all lifetime occurrences of infectious diseases in the average human being. Among the best-known infections caused by viruses are influenza, the common cold, measles, mumps, different forms of hepatitis, chicken pox, cold sores, genital herpes, and—as demonstrated in the 1980s—AIDS. Although not understood at the time I was starting my career in virology, virus infections are also thought to be responsible for about one-fifth of all human cancers, often by promoting chronic inflammation that eventually leads to cancerous growth.

Although the roots of virology go back to the nineteenth century, when Louis Pasteur investigated the origin of rabies and speculated that the disease might be caused by an agent too small to be seen under an ordinary light microscope (it can be visualized by electron microscopy, invented in the 1930s), the scientific concept of a virus as a submicroscopic infectious agent that requires living cells for its reproduction was not formulated until the 1920s. The golden age of virus research had only begun in the early 1950s, and I was entering the field at a fortuitous time as it was being firmly established.

Since I had no prior experience with laboratory techniques used in virus research, I spent my first two months at the Institute of Virology observing other scientists carrying out experiments. On completing my short apprenticeship, I was given a small laboratory of my own in the department of Dr. Helena Libíková, one of the established investigators at the institute who was focusing her studies on tick-borne encephalitis virus, or TBE virus for short. As the name implies, this virus is transmitted by deer ticks and causes a serious disease indigenous to Central and Eastern Europe, whose main characteristic is an inflammation of the brain. The disease can be fatal, and a significant percentage of patients who survive the infection end up with permanent neurological damage. Soon, I was assigned a full-time laboratory technician and a project. The goal of my project was to develop a tissue culture test to demonstrate the presence of TBE virus.

Let me explain what that means. In order to detect and identify a virus, for example from a sick patient, one needs a method that makes it possible to propagate the virus. Since viruses grow only inside living cells, the virus-containing material has to be inoculated either into a laboratory animal or into live cells grown artificially in test tubes. When I was beginning this project, it was known that inoculation of TBE virus into laboratory mice would cause an infection of the brain—encephalitis—that would eventually kill the mouse. However, inoculating mice is laborious, and there is a danger that the virus can be transmitted from the animals to laboratory personnel. In fact, several colleagues at the Institute of Virology contracted the disease while working with the virus and some ended up with serious neurological damage. A simpler system relying on test tube–grown cells would be safer.

Most viruses that cause diseases in humans or animals can be propagated in laboratory-grown cells called tissue cultures. Usually, when a virus multiplies in a tissue culture, the cells become sick and eventually die—a process that can be observed under an ordinary microscope. To the investigator, the appearance of a characteristic change in cell morphology serves as an indication of the presence of a virus.

TBE virus multiplied readily in many different types of cultured cells. The problem was that the virus—unlike many other viruses—did not produce visible damage to the cells, so that the virus-infected cells under the microscope were indistinguishable in their shape and appearance from the uninfected cells. Without a morphological change visible under the microscope, it was impossible to tell whether cells were infected with

TBE virus or not. In my search for a cell culture system in which TBE would produce a "cytopathic effect"—meaning a change in the appearance of the cells caused by the harmful effect of the virus—I tried a variety of test tube–grown cells under different conditions. Nothing worked. I became increasingly disillusioned and frustrated.

Frustration with the lack of progress in my laboratory work almost drove me to quitting research and becoming a medical practitioner. About a year-and-a-half after joining the Institute of Virology I thought to myself, "Why am I wasting my time here? I have a medical degree, I could be doing something productive." I contacted the head of the Department of Internal Medicine, whom I remembered from my medical student days. I explained the frustration with my laboratory research and asked if there might be a clinical trainee position available for me in his department. But it was the middle of the academic year and there were no openings. The head of the department advised me to come back in the spring when there might be an opening available.

I often think back to this moment and ask myself where I would be today if a position had been offered to me in the Department of Internal Medicine. Would I have started a career in clinical academic medicine? Become a practicing internist? I did not have the chance to find answers to these questions. By spring the tide had turned, my experiments started to work, and I continued my work at the Institute of Virology.

———

The Institute of Virology had been established in the early 1950s, largely as the brainchild of its founder and director, Dionýz Blaškovič. In the majority of countries, most scientific research is conducted within the walls of universities, but this was not the case in Czechoslovakia and other Communist countries. The Institute of Virology was part of the Czechoslovak Academy of Sciences, a structure built in Communist Czechoslovakia along the lines of a similar organization existing in the Soviet Union. After all, the Soviet Union was our "shining example," or so the official slogan of the time would have had us believe.

In the immediate post–World War II years, Blaškovič studied virology in the US with Jonas Salk, a prominent physician-scientist who—like Albert Sabin—later played a key role in the development of a vaccine against polio. Upon returning home, Blaškovič's ambition was to build an institute that would be highly regarded, not only in Czechoslovakia and the Communist bloc, but also in the West. He was a good organizer and a skilled politician, and to accomplish his goals he hired the best young scientific talent he could find in Czechoslovakia. Indeed, a large proportion of scientists recruited to the institute were from the Czech lands, not Slovakia—an unusual feature among establishments in Slovakia in those days.

But Blaškovič was also a pragmatist and he knew that he needed the support of the Communist authorities. Not only did he personally join the Communist Party, but in order to protect himself he also hired a handful of people with mediocre scientific credentials who had close connections to the Central Committee of the Communist Party. These highly visible Communist staff members served as a shield that protected the majority of his non-Communist scientists, like me, from excessive political scrutiny by the authorities. As a result, most of us were able to devote ourselves to professional work without having to worry whether our loyalty to the ideals of the "peoples' democracy" was under constant surveillance.

The science of virology was just beginning to take off; almost nothing was known yet about the molecular details of the structure of viruses and their mode of reproduction. There was so much to discover. We had the feeling, perhaps rightly so, that the quality of our research at the institute was not too far behind that of our colleagues in Western Europe or even in the US. What did hamper our work was the lack of some specialized equipment, and, in particular, shortages of reagents that had to be imported from Western countries.

Then there were the idiosyncrasies of working within the Communist system, such as the insistence that scientists should report to work at seven a.m.—the same time as factory workers—or that every morning we were required to attend a compulsory meeting (referred to as *desatminútovka*, literally a ten-minute event) devoted to the review of domestic and world political news; naturally, their interpretation had to reflect the official views. There were also compulsory meetings of the official trade union organization and of the ubiquitous Society for Czechoslovak-Soviet Friendship. We learned to keep our mouths shut in front of the few Communist fanatics and people who we suspected of being secret police informers.

These unpleasant elements were outweighed by the friendly, collegial spirit, especially among the younger staff members. We would passionately debate scientific problems, often long into the night. The most enthusiastic participant in these debates was Jan "Honza" Závada, a native of Prague, who at the end of the official working day would exhort me and a handful of other colleagues, "Come, let's chat about viruses." During these chats we let our imaginations loose as we tried to come up with ideas on how we could learn more about the nature of viruses as well as devise means to prevent and treat virus infections. Honza peppered the discussions with clever observations that he would introduce with, "What if . . ." As in, "What if this or that virus could be broken down into its constituents and then mixed with parts of some other virus to create a completely new virus?"—I am oversimplifying, of course. It was stimulating and a lot of fun!

———

After I graduated from medical school, I continued living with my parents in their apartment. This was a very common arrangement in those days, necessitated by the extreme shortage of housing. So scarce were rentable apartments that when young couples got married they almost inevitably had to continue living with one set of parents. There were waiting lists, and it could easily take five years or more for newlyweds to be assigned a government-owned rental. Single people were not even eligible to apply, and those who were from out of town could at best hope to rent a room in a shared apartment. My own situation was quite comfortable. In my parent's apartment I had a spacious room, with separate access from the entrance hall, which meant that I had some privacy and I could come and go without inconveniencing my parents. At the time my parents were not awash in money, so I contributed a portion of my salary to pay for my "room and board."

The nice plot of land overlooking Bratislava, where my parents had hoped to build a family villa, had been expropriated during the war. The new Aryan owners built a stone wall along the front of the property and a garage, but they ran out of time to build a house. Immediately after the war, my parents reclaimed the property, but then the Communist take-over once more thwarted my parents' plans to build a family home. The land was small enough not to be confiscated. Larger tracts of land and apartment buildings with multiple dwellings were "nationalized," while ownership of smaller plots, family gardens, and single-family homes was tolerated throughout the Communist era. Even though no one in our family was fond of gardening, we used the piece of land to plant vege-tables and strawberries. There were also a dozen or so apricot and plum trees, and when the season was good, I would go to the farmers' market to sell our fruit in order to make a few extra crowns.

As I neared the completion of medical school, I got too busy to tend to the garden, and my parents just got plain tired of the work. So we made the decision to sell the land to raise some cash and use it to make our lives more comfortable.

By the end of the 1950s, the regime—though certainly not turning into a democracy—was beginning to show signs of a slight political thaw. Stalin died in 1953. Nikita Khrushchev succeeded him as first secretary of the Communist Party, later becoming premier of the Soviet Union. Khrushchev initiated a gradual process of de-Stalinization culminating in his famous secret speech at the closed session of the Twentieth Con-gress of the Communist Party of the Soviet Union in 1956. Echoes of the de-Stalinization process reverberated in the satellite countries.

Accompanying the very slow and not always straightforward process of de-Stalinization was the emergence of a modest improvement in the standard of living, visible especially in the economically more advanced East Bloc countries including Czechoslovakia and East Germany. Some people in Czechoslovakia—still a very limited number—became moderately prosperous, at least when compared to the early 1950s. The people who lived a little better than the average citizen were a handful of artists and writers supported by the government, some professionals, especially physicians who accepted payments under the table, and individuals or families with relatives in the West who sent gifts of Western currency that could be used to buy special goods available only in officially sanctioned government-owned *Tuzex* stores. One sign of a modicum of newly found prosperity was that a few people were able to afford to buy a private car, something that was unheard-of in the early 1950s.

When my parents sold their piece of land they suddenly ended up with a significant amount of cash—a sum that in those days corresponded to about four annual salaries. Almost everyone—factory workers and professionals alike—were paid about the same meager wages. With very few exceptions, the ratio between the lowest salary of an unqualified worker and that of a director of a large enterprise was perhaps one to three. Holding on to cash was not considered a good idea lest the government introduce another forced "monetary reform" that could result in the loss of a significant chunk of one's savings. So we decided to spend the cash on something concrete.

My parents (with my active encouragement) decided to put their name on a waiting list for the purchase of a car, which cost about the same amount they netted from the sale of the land. After paying a deposit, the wait was around three years. In preparation for the day when we would be able to take possession of the car, both my mother and I took driving lessons and passed the required examination and road test. At the age of sixty-something, my father was not interested in learning how to drive. Then, one day in 1960, we were called to take possession of the car—a small, rather cheaply constructed Czech-made Škoda two-door coupe. It turned out that despite obtaining her driver's license, my mother was too scared to drive the car, and I became the sole driver and de facto owner. This was to be the car in which I and my future wife, Marica, drove off into our new life.

During my years at the Bratislava Institute of Virology we quite frequently welcomed visiting scientists from Western countries. Foreign visitors would meet with members of the staff, including novices like myself. Parenthetically, meeting visitors from the West provided not only an opportunity to learn about the latest advances in virology, it also offered a chance to obtain firsthand information about life on the other side of the Iron Curtain—a term popularized by Winston Churchill in his 1946 speech that denounced the creation of a Soviet-dominated bloc of Eastern European countries. Though the Iron Curtain, marking the border between Czechoslovakia and Austria, was located right across the Danube, merely a few miles away from the center of Bratislava, crossing it was virtually impossible for the vast majority of citizens.

Together with some of my colleagues, I often volunteered to entertain foreign visitors after working hours. I don't think our relatively prosperous Western visitors realized that we had no expense account and paid for all the entertainment from our own meager resources. Our salaries were indeed quite low, and we would regularly run out of cash before the end of each month, often borrowing small amounts of money from our more disciplined friends or parents to tide us over till our next payday. Yet the money we spent entertaining Western visitors was a good investment, well worth the financial sacrifice, because of the contacts we made and the opportunity to broaden our horizons.

One visitor who very much influenced the course of my work was Alick Isaacs, a prominent British virologist. Alick came to Bratislava in 1957, the same year he published his first papers on the newly identified protein called interferon, authored jointly with his Swiss coworker, Jean Lindenmann. This also happened to be the year I graduated from medical school and joined the Institute of Virology.

I still remember Alick's lecture in Bratislava and his vivid description of interferon. The full significance of his discovery would not become appreciated until many years later. Had Alick not died at a young age, he almost certainly would have received the Nobel Prize.

At the time I met him in Bratislava, Alick, though already well known among virologists for his studies on influenza virus, was a youthful-looking man, bubbling with energy, wit, and personal charm. Unlike many other visitors from the West, Alick did not hesitate to ask pointed questions unrelated to science. When the two of us were out of other people's earshot he inquired how I felt about the Communist system. I did not conceal my lack of enthusiasm for the "workers' paradise."

Another relatively frequent visitor to the Bratislava Institute of Virology was Albert Sabin, the American scientist famous for developing the live oral polio vaccine. The Soviet Union and Czechoslovakia were among the first countries to adopt the Sabin vaccine. I remember especially well Albert's visit to our Institute in 1959. During the thirty minutes or so I spent with him, I told him about my unsuccessful struggle with the development of a cell culture–based assay for TBE virus. Could he suggest some trick that would prod TBE virus into producing a visible morphological change in cell cultures?

Albert advised me to try a different approach. Rather than attempting in vain to find ways to get the cells damaged or killed by the virus, I could inoculate cultures with TBE virus, wait two to three days, and then infect the same cells with another virus that is known to kill cells. Chances are, Albert said, TBE virus will induce a state of interference so that the second virus will *fail* to kill the cells. The presence of TBE virus could then be inferred from the absence of cell killing. Albert mentioned that a similar strategy had been used for the detection of some other viruses that, like TBE virus, would not damage and kill infected cells.

I tried the strategy suggested by Albert Sabin. My technician and I inoculated cells derived from chick embryos with TBE virus and, after waiting two or three days, we added another virus, the Western equine encephalomyelitis (WEE) virus. The experiment was instantly successful! One day after WEE virus inoculation, cells infected with WEE virus alone were killed, but cells inoculated first with TBE and then with WEE virus looked healthy. As Albert Sabin had predicted, TBE interfered with the multiplication of WEE virus and protected cells from killing caused by WEE virus. After only a few more weeks of work I wrote my first independent paper entitled "Interference between tick-borne encephalitis virus and Western equine encephalomyelitis virus in chick embryo tissue cultures," published in 1960.

I did not know—nor would I have appreciated the knowledge at the time—that Albert was a graduate of New York University School of Medicine in New York City, where, some six years after our conversation, I would join the faculty, and where I would spend no less than fifty years. Forty-five years after my brief, illuminating conversation with Albert, I endowed the Albert B. Sabin Professorship at NYU School of Medicine to honor the memory of the great scientist who played such an important role in my own scientific beginning.

I don't remember when it first occurred to me that a factor similar to Alick Isaacs's interferon could be responsible for the inhibition of WEE virus multiplication by TBE virus in my chick embryo cell cultures. Why did I not think of this possibility right away? At the time, in mid-1959, the role of interferon in virus interference was not yet widely recognized. Alick himself initially considered interferon to be a specific product of cells exposed to influenza virus. That most viruses could induce the synthesis of interferon in a wide variety of cells was not even suspected.

I started to think about the mechanism of the interference between TBE and WEE viruses toward the end of 1959, as I was writing my paper describing the application of this interference phenomenon for the assay of TBE virus. One observation I made suggested that TBE virus itself was not responsible for the establishment of interference. I found that it took two to three days from the time of inoculation with TBE virus for the cells to become resistant to WEE virus. Yet fluids from TBE-infected cultures, when transferred onto fresh chick cells, induced resistance to WEE virus much faster—within hours.

To prove more directly that interference was produced by a factor different from TBE virus, I collected fluids of infected chick embryo cell cultures that contained TBE virus, then got rid of the virus by the addition of specific antibodies that inactivated the virus. Inactivation of the virus did not diminish the ability of the culture fluids to make cells resistant to inoculation with WEE virus. Clearly, something produced by the cells in response to TBE virus, but not TBE virus itself, mediated the interference. Was it interferon?

I can still recall my excitement when I realized that I was in pursuit of a new and interesting story. I hardly slept for several weeks. When I thought I had enough data to write another paper I sat down and prepared the first draft in only two days. Then, after some additional polishing, I took the manuscript to Blaškovič, the director of the Institute, and told him I wanted to submit it to the prestigious British journal *Nature*.

A few days later Blaškovič called me to his office. "I like your paper," he said, "but instead of submitting it to *Nature*, wouldn't it be more appropriate to publish it in *Acta Virologica*, our own journal?" *Acta Virologica* was a journal edited by Blaškovič and printed right there in Bratislava. Well, I didn't want to publish the article in a local journal where it would be much less likely noticed by the international science community. I no longer remember how I convinced Blaškovič to let me send the paper to *Nature*. In any case, some months later, in July of 1960, *Nature* printed the report, titled "An interferon-like substance released from tick-borne encephalitis virus-infected chick embryo fibroblast cells." The article contained two tables, one figure, a total of only four references, and a byline listing the name of a single author—me.

This is how a few weeks after my twenty-seventh birthday and only three years after graduation from medical school, I had my own paper published in *Nature*—then, as now, considered the most prominent international science journal. Such were the romantic days of biomedical research, when common sense and a little good luck were more important than access to the latest technological advances. I should add that in the US I would not have been able to publish a paper at this early stage of my career without including my mentor as coauthor. But in the anomalous situation of Czechoslovakia such rules did not apply.

The work published in *Nature* also served as the nucleus for my PhD thesis. I had enrolled in the PhD program in 1958 (then still known under the Soviet-inspired name "Candidate of Sciences" program), thinking that my thesis would deal with the assay of TBE virus. So new was the field of interferon research that my dissertation, which I defended in 1962, earned the distinction of being the very first PhD dissertation in the world devoted to interferon. Thousands of PhD degrees have since been granted to students working in this field.

Chapter Eight

Like a Country Vicar in Rome

My first interferon publication in 1960 was followed by many other studies, some authored jointly with colleagues from the Bratislava Institute of Virology. Over the years I have enjoyed playing the role of a scientific iconoclast. According to the then-current dogma, mice that recover from infection with Sindbis virus, another encephalitis-causing virus, should have had higher levels of interferon in their brains than mice that succumb to infection, but what I found was the exact opposite. My findings were printed in *Virology*, a widely circulated journal published in the US. Later studies by Ion Gresser and other investigators demonstrated that interferon production, while protective and beneficial under many circumstances, could also be harmful.

Another study done jointly with my first and only graduate student in Bratislava, Daniel Stanček, showed that even though interferon could protect cells from an ensuing TBE virus infection, the addition of interferon after cells had become fully infected failed to suppress TBE virus multiplication. This was an early demonstration of the now widely known fact that in order to become successful pathogens, many viruses developed the capacity to counteract the ability of interferon to inhibit virus infections.

These publications led to many contacts with scientists in other parts of the world who had begun to study interferon at about the same time. Apart from Alick Isaacs's laboratory in London, an active center of interferon research emerged in Belgium at the Catholic University of Leuven. Jean Lindenmann, coauthor of the original publications describing interferon, had returned from Alick's lab in London to the University of Zurich, and though not planning to continue interferon research, he soon reemerged as an active contributor to the field.

Other laboratories active in interferon research in the early 1960s were at Harvard University Medical School in Boston, the Children's Hospital in Philadelphia, Johns Hopkins University in Baltimore, the National Institutes of Health in Bethesda, Cornell University Medical College in New York City, and in Villejuif, outside Paris. I exchanged correspondence with many colleagues from these institutions and some, including Jean Lindenmann, and Robert Wagner from Johns Hopkins University, I met in person during their visits to the Institute of Virology in Bratislava. A field of interferon research, somewhat separate from the rest of virology, was gradually forming.

In the fall of 1962, the authorities granted me a two-week study trip to Belgium and England. The barrier between East and West in those days was formidable. There were only two passengers on the spacious Czechoslovak Airlines plane during my trip from Prague to Brussels: an employee of the Czech embassy and me. My first stop was Leuven (then still known to the outside world as Louvain). The charismatic director of the Rega Institute, Pieter De Somer, already in those days had the idea— or shall we say, dream—to develop interferon production commercially. In this and other ways Pieter De Somer was clearly ahead of his time.

In Leuven I stayed at the house of Edward De Maeyer and Jaqueline De Maeyer-Guignard—marking the beginning of a wonderful friendship between us that flourished till their passing some years ago. At the time, the De Maeyers were relatively new to the Rega Institute, having returned to Belgium as newlyweds after completing their fellowships in the United States. Edward became interested in interferon during the final stage of his sojourn at Harvard University in the laboratory of John Enders, a Nobel Prize winner and pioneer of modern virological research. In 1962, Edward was already a driving force on De Somer's interferon team. His Swiss-French wife, Jaqueline, trained as a pediatric endocrinologist, was now working alongside Edward in interferon research.

From Belgium I continued on to London where I spent several exciting days at the National Institute for Medical Research in Mill Hill on the outskirts of London. Most memorable was the time spent with Alick Isaacs, whom I had not seen since his visit in Bratislava in 1957. I was introduced to Alick's colleagues whose names I knew from publications, including Derek Burke, Joyce Taylor, David Tyrrell, and Tony Allison. Being in a major metropolis in the free world for the first time was an overwhelming experience. This is what I said as an introduction to my seminar at Mill Hill: "I feel like a country vicar who comes to Rome to lecture about the Bible to the pope and his cardinals."

At the end of my stay in London, while saying goodbye to me, Alick asked whether I would be interested in spending a year in his laboratory as a visiting fellow. They had some funds, he said, to pay me a stipend. This was a fantastic offer and I had great difficulty suppressing my excitement. The Mill Hill institute was a major center of medical research and Alick was a leading figure in the world of virology. In Bratislava I was the only person focusing on interferon and there were few people I could ask for advice. In London I would have the opportunity to collaborate with and learn from many others working in the field. It would be a life-changing experience!

But right away I realized that it would be hard to obtain approval from the authorities in Czechoslovakia. There were some other people at the Institute of Virology who had been permitted to spend up to one year in Western countries, but all of them were more senior than I. The fact that I was not a member of the Communist Party and—in the official parlance—"politically passive" would not be helpful either.

My fears were well-founded. Director Blaškovič was mildly supportive, but said right away that my stay abroad would have to be sanctioned by the local branch of the Communist Party. The denial of my request arrived rather promptly. I was told that there were other comrades at the institute more senior and more in need of advanced training, and they had to be given priority. The Communist Party bosses ignored the fact that I was the only person offered the opportunity to spend time in Isaacs's laboratory and no one else could take my place there.

When I returned to Bratislava from my short trip to Belgium and the UK, I suggested to Director Blaškovič that we organize an international meeting devoted to interferon. Interferon research was just beginning to emerge as a separate discipline, but I was confident that with my newly established contacts we could attract enough participants to what was to be the first international gathering of interferon experts. Blaškovič, who always sought greater visibility for himself and his institute, liked the idea.

And so, in September 1964, the Institute of Virology sponsored the first international conference on interferon, held near Bratislava at the Smolenice Castle, an imposing neo-Gothic edifice built just before the outbreak of World War I by Count Pálffy, a wealthy Hungarian aristocrat, as a family residence; it was later confiscated by the Czechoslovak government and converted into a conference center for the Academy of Sciences. I acted as the scientific secretary of the conference, and was instrumental in the planning of the sessions.

Interferon research was still a very small field and probably half of the world's then-active investigators participated. Alick Isaacs was scheduled to give the opening lecture. Unfortunately, Alick became ill and could not come. (His illness marked the beginning of a string of serious health problems that ended with his premature death a few years later.) However, two of his colleagues from Mill Hill, Tony Allison and Joseph Sonnabend, did participate. The very active interferon group from the University of Leuven was represented by Pieter De Somer, Edward De Maeyer, and Jaqueline De Maeyer-Guignard. Participants from the United States included Sam Baron, Robert Friedman, Kurt Paucker, Thomas Merigan, and Edwin Kilbourne—all to become prominent in interferon research. From France came André Lwoff, the Nobel Prize winner who was then interested in host defense mechanisms to virus infections. Another participant was Kari Cantell from Finland, who would later become a pioneer in the clinical use of interferon. There were many attendees from Czechoslovakia, the Soviet Union, and other Communist bloc countries.

One evening, conference participants gathered to play a social game. A moment that has stayed with me in particular was when Pieter De Somer (later to become president of Leuven University) answered the question: "What would you do if you were to discover an effective cure for virus infections?" "I would sell it," he said without hesitation. This moment may well have marked the conceptual birth of commercial biotechnology as we know it today.

An unavoidable fact of life in all East Bloc countries was the ubiquitous presence of the secret police—KGB in the Soviet Union, Stasi in East Germany, and Štátna bezpečnosť or ŠtB (pronounced *esh-tay-bay*) in Czechoslovakia.

I had my first serious encounter with the ŠtB when I was a medical student. In her ophthalmology office, my mother worked with a nurse whose husband happened to be a ŠtB operative. One day the nurse told my mother that her husband's organization was keeping an eye on my mother's internist colleague Dr. Bučko, who had his office in the same ambulatory care center as my mother, because Dr. Bučko often made remarks that were hostile to the Socialist system. My mother casually mentioned this conversation to me over the dinner table. Dr. Bučko's son, nicknamed Fero, happened to be my classmate in medical school. Without informing my mother, the next day when I saw Fero I told him what I had learned from my mother and suggested that he advise his father to be more careful about what he was saying in public about the regime.

I am not sure what happened next, but perhaps Dr. Bučko, instead of heeding my advice, went to complain to the authorities that they were spreading false rumors about him. In any case, I soon received a subpoena to appear at the notorious building housing the ŠtB headquarters on the outskirts of Bratislava. The ŠtB agent who interrogated me was not friendly. He wanted to know exactly what I had said to my classmate and why and how and when I received the information. I was forced to reveal that I heard it first from my mother. My mother was called in next and asked similar questions. Afterward, they called me in once again. With a stern expression on his face, the ŠtB agent told me that they were letting me go this time, but should I be called in again, things would become much more serious.

Another encounter with the ŠtB happened shortly after the defection of my friend Bobby Lamberg, when I was already working at the Institute of Virology. Once again I received a subpoena, although this time I was to appear not at the ŠtB headquarters but at the main police office building—which happened to be the same building where my father was jailed for a few days during World War II. I had no idea why I was called in, but as I was waiting outside the office a young man came out whom I recognized as a close friend of Bobby's. I was called in next.

There were two men interrogating me. I could truthfully tell them that I knew nothing about Bobby's plans to defect. It would be rare even for very close friends to discuss plans of defection. It was too dangerous for two reasons. First, if the plans were to somehow leak out, the would-be defector would end up in jail instead of in exile. Second, when friends who stayed behind were questioned by the ŠtB, not knowing about the plans was safer. The ŠtB agents kept asking me questions about how closely I knew Bobby, about his political views, and about what other friends of his I knew. They were mildly intimidating. Fortunately Bobby had moved to Prague some years earlier, and I could say that I had not had close contact with him since then. (This was not the complete truth because I would always meet Bobby during my frequent visits to Prague.)

I also remember that I was called in by the ŠtB after my return from the trip to Belgium and the UK. The person who met me introduced himself as "Mráz" ("Frost" in Slovak), which almost certainly was not his real last name, but it matched his glacial expression. I was first reminded of how privileged I was to have received the permission to travel to the West. (The clear implication was, you'd better be cooperative or else you will not be allowed to travel again.) They asked me for the names of everyone I met during my trip and what their positions were. Did I discuss politics with anyone? (I did not, I said.) Did I meet any Czech or Slovak defectors? (I did not.) Did anyone try to talk me into defecting or recruit me to become a spy for the West? (Nobody did.)

Some months later I was confronted with an even more difficult situation. I was called to the ŠtB and met once again by Mr. Mráz. He started out in a friendly tone, telling me that he knew I was a loyal citizen and a respected scientist. He went on to note that I was fluent in English, and that he wanted to ask of me a favor. A scientist from the US was scheduled to visit another Academy Institute. They would arrange for me to be introduced to the visiting scientist from America, and I would then be expected to spend some time engaging him in a conversation and perhaps invite him for a drink. At some point I would introduce him to another person designated by Mr. Mráz, and this other person would then invite the American scientist to join him on a deer hunt. That was all I would have to do, said Mráz.

As I listened to Mr. Mráz explaining his plan, I became increasingly alarmed. I knew that refusing to cooperate could come at a high cost. At the same time, I realized that if I did what Mráz asked me to do I would be forever obligated to carry out any dirty work the ŠtB asked of me. "Look," I said, gathering all the strength I could muster, "I really cannot do what you are asking me to do. The American scientist would quickly figure out that I am doing this for the secret police and my reputation as a scientist would be ruined. I think I can be more useful to our country by continuing to do decent scientific work." To my surprise, Mráz did not insist. I was fortunate, but how much longer could I resist entreaties by the ŠtB?

———

There was another, strictly confidential item on my agenda at the Smolenice interferon meeting. By now I was married. Some weeks before the meeting, my wife Marica and I had played hosts to a colleague from Vienna, professor Hans Moritsch, head of the Department of Hygiene and Microbiology at the Vienna Medical School, and his wife Edda, who stayed with us in our home in Bratislava and then extended an invitation for us to spend a weekend with them in Vienna. We explained that it was unlikely the authorities would grant us permission. In the rare instances when someone was allowed to travel to a Western country, common practice in the Eastern Bloc was to ensure that a spouse or a child was kept behind so as to make defection unlikely. As we had no children, we thought the secret police would consider it too risky to permit Marica and me to visit Vienna together.

Nevertheless, we asked the Moritsches to mail us a written invitation so that we could submit an application and hope for a miracle. The invitation from the Moritsches to visit them for a weekend arrived, and a few days before the opening of the Smolenice conference, Marica and I submitted our applications for a travel permit.

There was something we did not mention to the Moritsches at the time. Ever since getting married two years earlier, Marica and I often discussed the idea of moving to the US where Marica's older brother lived and worked as a physician. Since getting a permission to emigrate from Czechoslovakia to a Western country was impossible for people of our generation, we came to an agreement that if the opportunity arose, we would defect.

I knew that if we defected, I would need help to find a job. I had to decide who among the Western colleagues I knew would be trustworthy enough to be told in confidence about our plans. As the conference approached, I decided that it would be Edward and Jaqueline De Maeyer, with whom I had developed a friendship during my visit to Leuven when I had stayed in their house.

Since I suspected that the walls of the Smolenice Castle were bugged, I told Edward and Jaqueline about our secret plans during a walk in the surrounding gardens. If Marica and I were to get to Vienna, would they be in a position to find me a temporary position at their institute in Leuven? We knew that if we were able to get out of the country, we would want to settle in the US, but we had no idea how long it would take us to get American immigration visas, and we would need a home in the interim. Edward offered to consult in confidence with their influential director, De Somer, who happened to be in Smolenice too.

Soon Edward and I took another walk through the gardens during which he told me that De Somer promised to do his best for us, but he needed more time to determine if he could find the necessary resources to offer me a job. Okay, but how would Edward let me know whether I could count on getting a position in Leuven? We came up with our own John le Carré–style code: If De Somer told Edward that there was a position available for me, Edward would mail me a preprinted postcard requesting a copy of my latest scientific publication. In the pre-Internet days such "reprint request cards" were being routinely exchanged among scientists all over the world, and addressing one to me would not arouse suspicion.

Then, in late September, we received the official permit authorizing Marica and me to travel to Austria. We were overcome with conflicting emotions: the great excitement that our desire to move to the West might become a reality, mixed with more somber thoughts having to do with leaving behind our families and friends and not knowing what we would be leaving them for. Would we have the courage to go through with it?

PART FOUR | # In Pursuit of a New Life

Chapter Nine

Marriage and Defection

Marica Gerháth and I were introduced by my long-standing friend Šaci Vietor and his wife, Nina Vietor. Marica and Nina were classmates in high school and continued their friendship at the university where Marica studied art history, and Nina majored in English and Russian. Šaci and Nina invited Marica and me to their house around Easter in 1961 for a meal featuring a traditional Eastern European delicacy called *paskha*, made from sweetened cottage cheese.

Although Marica's birth certificate spells her first name "Mária," her parents and everyone else always called her "Marica." The name is said to have been inspired by Emmerich Kálmán's operetta *Gräfin Mariza* (Countess Maritza), transcribed as "Marica" in some European languages, including Slovak, Czech, and Hungarian.

To this day, neither Marica nor I are sure if the invitation was intended as an exercise in matchmaking or whether it just happened that both of us were invited at the same time. Of course, I took notice of Marica—she was a woman who would turn the heads of strangers in the street. At almost five feet eight inches, she was unusually tall. She also dressed in a way that, though not ostentatious, was distinct from how other young women in Bratislava dressed in those days. Most of Marica's clothes were gifts sent by her older brother, a physician who defected from Czechoslovakia in 1957 and by then lived in New York City. Because I had a general interest in the arts, I was impressed, too, with Marica's position as an assistant curator at the Slovak National Gallery. Yet I did not ask Marica out, perhaps because she was quite different—certainly less coquettish— than the girls I was casually dating at the time.

In the summer I departed for a three-month-long professional visit to

Moscow, returning in early November. One late afternoon soon after my return from the Soviet Union, I decided to go see the art on exhibit at the Slovak National Gallery and there I ran into Marica who was working with a colleague on the installation of an exhibition. We said hello, exchanged a few words, and this time I did ask her for a date.

We formed a friendship that rapidly progressed to a serious romantic relationship—I will admit that it took me a little while to get used to Marica being about a half-inch taller than me. I could not have imagined forming a lifelong relationship with any of my former girlfriends but, to my surprise, the thought of spending the rest of my life with Marica occurred to me relatively soon—even though our backgrounds were distinctly different.

Not only were our professional fields unsimilar, so were our family roots. Marica was raised Catholic, and both her mother (who had died of breast cancer when Marica was a teenager) and her father were school-teachers. While my parents were quite worldly in their views and life-style, her father was more conservative and less open to nontraditional influences. Marica's family was well-to-do until the Communist take-over, because her father owned farmland, but their lifestyle had never been flamboyant. The Gerháths were much less affected by the pro-Nazi regime during the Second World War than my family, but because they were landowners, they suffered more material losses and deprivation under the Communists.

I felt comfortable and relaxed in Marica's company. We found that we had many common interests and our tastes in art, music, even food, were quite similar. Our dislikes were similar too. We hated the govern-ment-promoted art of Socialist realism and everything that had to do with Communist ideology and propaganda. We both despised pretense and shared a youthful disrespect for conventions.

My parents liked Marica too. My father, always friendly and open, immediately formed a cordial relationship with Marica. My mother took a little longer to warm to her, but soon she too developed an appreciation of Marica's quiet and considerate personality. It helped that Marica genu-inely admired my mother's colorful persona, and tolerated her occasional eccentricities.

By the spring of 1962, a tacit understanding had developed in my family that Marica and I would form a permanent relationship. As for Marica's father, I had met him during my visits to their house. And though I was not sure, I had a feeling that he did not dislike me and probably would not mind if I were to ask for his daughter's hand.

I don't remember formally proposing marriage. Both Marica and I felt strongly that we didn't want to get tied down by tradition. A formal engagement with rings, followed by a big formal wedding ceremony, was not something we ever contemplated. Perhaps we were a reflection of the period, when lives were much less settled than in previous times. Even though some young couples still arranged big weddings with formal gowns and all that entailed, this was never something that interested us.

In early July, my parents left for a three-week vacation in Mariánské Lázně (formerly Marienbad), a spa in western Czechoslovakia. Marica and I were scheduled to visit my parents there toward the end of their vacation and spend a few days with them. I am not sure when exactly Marica and I came up with the idea to have a small civil wedding ceremony during my parents' absence. We completed the necessary formalities on short notice.

The night before the scheduled wedding, during dinner at their house with Marica, I told Marica's father that we had decided to get married. He seemed surprised, but in his characteristic manner did not show much emotion. He asked when we wanted to have the wedding. I said "soon," and made it clear that we didn't expect him to organize a wedding reception. Marica had an unusual grin on her face while I had this exchange with her father, which I read as something like, "Okay, you sweat it out, I'll stay out of this."

The day before the wedding we also informed our two chosen witnesses, Lida Peterajová, a colleague of Marica's at the Slovak National Gallery, and Břetislav "Slávek" Rada, my colleague at the Institute of Virology. Our prenoon wedding involved minimal ceremony. Marica was wearing a simple summer skirt with a blouse; I was dressed in a light summer suit with shirt and tie.

The county clerk who administered the wedding vows was heavyset and unattractive. "Do you want to exchange wedding rings?" he asked at one point. "We don't have wedding rings," Marica and I answered in unison. After the short ceremony we did have a nice lunch with Lida and Slávek at the Hotel Devín, adjacent to Marica's office at the Slovak National Gallery, whereupon we all returned to our respective workplaces.

The next day we drove off on our honeymoon. We had a few days before the date we had agreed to join my parents. Mariánské Lázně is a three-hundred-mile drive from Bratislava, so we decided to take a circuitous route, first visiting some medieval castles in northern Slovakia. Our last stop before Mariánské Lázně was in the city of Hradec Králové in northern Bohemia. The desk clerk at the hotel refused to let us share a room because our ID cards still listed us under two different last names. The argument was fixed with a small bribe.

The following day we arrived in Mariánské Lázně. During the journey Marica and I discussed how best to deliver the news to my parents. Our decision was to do it as soon as we saw them. But when we got there we found my parents anxious to leave for an outing. "Good that you are finally here," my mother said. "Get ready right away, we are all expected at the Hotel Esplanade where we'll be joining our friends for dinner." Marica and I looked at one another and tacitly decided that we'd have to postpone breaking the news to my parents.

By the time the dinner ended it was late, and we were not staying in the same place as my parents. When we finally did break the news to my parents at breakfast—we were all sitting down when it happened—they took it remarkably well. My father ordered champagne to congratulate us, and my mother joined in the toast. It took my mother a little longer to recover completely. First she said she was sorry we did not let her know earlier, because she would have liked to tell her friends about it. Later she admitted that doing what we did saved her work and aggravation.

We decided to move in with Marica's father. He lived in a handsome, albeit slightly rundown, two-story family residence built in the 1920s. My father-in-law continued living on the main floor, and we moved to the three rooms on the second floor, one of which also served as a kitchen. As was customary in those days, there was only one bathroom in the house that we shared with Marica's father. Most evenings Marica, with my inexpert help, cooked dinner for the three of us. Our Sunday routine was to be my father's guests for lunch at a restaurant (without my mother, who claimed to be on a diet), after which we would go to my parents' house for dessert and conversation.

Not only were Marica's and my backgrounds quite different, but our political views, too, had been formed somewhat differently by our circumstances. While I disliked the Communists, Marica's aversion to the regime was deeper than mine, because her family's life was more strongly impacted by the political system. Originally a schoolteacher, her father had been promoted to an administrative position at the Ministry of Education. After the Communist takeover he was fired from the job at the ministry and assigned to a petty clerical position at the university library, suffering a cut in his income. Marica's parents had a difficult time adjusting to the resulting hardships.

Marica's mother, who was a headmistress at an elementary school, developed breast cancer when Marica was in her early teens and died a few years later. At age seventeen, while still at school, Marica had to take on responsibility for the running of the household, including the care of her brother Pat'o, nine years her junior.

Then, when Marica was at the university, her older brother, Ivan, defected to the West. Miraculously, she was not barred from continuing her university studies, but when she graduated, her references, prepared by the omnipresent "department of cadres" at the university, described her as politically unreliable, and she was, at first, unable to get a job. When, with the help of a sympathetic colleague who happened to be a Communist Party member, she did get her position at the National Gallery, she had faced bias. Not only was she never allowed to travel to any Western country, she was also dropped from the list of participants for a visit to the Soviet Union.

Marica made it clear to me that if she had the opportunity she would leave Czechoslovakia for good without shedding a tear of regret. Because I saw no realistic way of how the two of us might be able to defect, I tried to persuade her that we could live a relatively happy life in Czechoslovakia. She was not convinced. "I am too tall for this country," she would say, only half-jokingly.

Her older brother Ivan was writing enthusiastic letters about his life in New York, and to prove the point he was sending Marica regular packages with generous quantities of clothes. Marica loved to dress up (a passion she has retained through today) and being able to dress well—certainly by Bratislava standards of those days—was one thing that kept her morale up.

The question of defection had often come up in our conversations. I too was frustrated with many aspects of life in Communist Czechoslovakia. Eventually I agreed that if the opportunity to leave ever arose, we would get out.

The opportunity did arise when in September 1964 the authorities unexpectedly granted us permission to pay a weekend visit to our friends Hans and Edda Moritsch in Vienna.

Marica, perhaps feeling guilty about abandoning her widowed elderly father and much younger brother, started having second thoughts. But this time I was firm. I reminded her about the many discussions we had in which she would tell me how much she longed to get out of Czechoslovakia. "You were always convincing me that we should leave if the opportunity arises, and now we have a realistic chance," I said.

As the invitation from the Moritsches was for a weekend stay in mid-October, the departure date was set for Saturday, October 10. An important decision to make was whether to tell our parents about the plans. Even though my parents often commented that we would be better off in the West, we had never before been in a situation to actually do something about it. I decided that I would let my parents in on the secret, and Marica agreed.

We debated whether we should tell Marica's father, but in the end we decided against it. One reason was that we lived in his house and undoubtedly Marica's father would be the first to be questioned by the ŠtB when we had not returned by October 12, our scheduled date to be back home. It would be in his interest to be able to tell the secret police truthfully that he knew absolutely nothing about our plans to defect. Another reason was that Marica's father was emotionally more fragile than my parents, and we could not predict what his reaction would be if we told him that we were leaving for good.

My parents were supportive of our plans. They said right away, "Go! You will have a better future in the West."

It was not difficult to imagine how hard this was for them. They—and we—had no idea how long it would be before we saw one another again, and there was the possibility we would not see each other again in our lifetimes. Other than my parents we told no one.

We did not worry much about the possessions we would be leaving behind. Marica and I did not own a great deal. We lived in the house that belonged to Marica's father. Most of the furnishings we had in the three small rooms we occupied were hand-me-downs from my parents and Marica's father. There were a few works of art received as gifts and a nineteenth-century writing desk acquired by Marica from money she put aside from her modest income. As was true of most of our friends, we lived from paycheck to paycheck, we had no personal savings, and, in any case, we were allowed to carry only a minuscule amount of cash to Vienna. Our most valuable material possession was the small Škoda automobile my parents had bought when they sold their parcel of land. We planned to travel to Vienna in our car, so that we could take it with us to Belgium or whatever other country might turn out to become our destination.

Concealing our emotions was an enormous challenge. At work, we had to pretend that all was as usual. It helped that I was working fever-ishly to complete the assembly and editing of manuscripts for the pro-ceedings of the Smolenice interferon meeting. (Although by the time of my departure all manuscripts were ready to be sent to the printer, the local powers that be decided that publishing the conference proceedings, be it with or without my name included, would be politically too embarrassing, given my defection.)

A few days before our scheduled departure my technician came running to let me know that I had a phone call from Vienna. I froze up. It was very unusual to get a phone call from a Western country, especially at work. Then I thought to myself, "It must be Hans Moritsch cancelling or postponing the invitation." I grabbed the phone, shaking. It was indeed Moritsch calling. He said, "I have tickets for the Vienna Opera for all of us. Can you bring your tuxedo?"

On October 10, 1964, about two weeks after receiving the permission to visit the Moritsches in Vienna, we packed two suitcases into our small Škoda automobile and departed for Austria.

The border between Czechoslovakia and Austria was part of the Iron Curtain—equipped with watchtowers, minefields, and electrified wire fences—separating Communist countries from the free world. Even before reaching the border, only a few miles removed from the center of Bratislava, we had to pass through a checkpoint manned by armed guards. At the border crossing we waited in trepidation as the Czechoslovakian border guards examined our papers, hesitating for the longest minutes of our lives before they let us pass to the other side. Would they become suspicious because we were carrying heavy winter coats for our three-day visit to Vienna in early October? Would they search the contents of our two bags and find that we had packed more than three shirts and three sets of underwear? The guards opened the car trunk to make sure we were not smuggling someone out of Czechoslovakia. They inspected the underside of our car. But they ignored the winter coats and they were not interested in how many shirts or pieces of underwear we were carrying.

Once on the right side of the border, the free Austrian side, we were elated. On a deserted two-lane highway about a half-mile from the border crossing, safely out of the reach of the Czech guards, Marica and

I stepped out of the car and embraced. "We made it," I said, my voice trembling. She nodded, wordless, her eyes glistening with emotion.

Now we could begin to shed the fears not only of the last two weeks, but also of the years before. Vienna was forty miles ahead of us. We could not begin to guess what lay beyond. We got back into the car and drove on.

"*Seid Ihr sicher?* Are you sure?" asked Hans and Edda Moritsch, with incredulous expressions. Yes, we were sure. Absolutely sure. We were not returning to Bratislava.

We had arrived at the Moritsches' house on the outskirts of Vienna only minutes earlier. Sensing they might be afraid that, instead of the planned two nights, we would now be staying with them indefinitely, we were quick to reassure them that we intended to move on very soon.

We told them Marica had a brother in the US whom we would be contacting shortly, and would borrow money from him to tide us over until we were able to get settled. We also told them we expected that I would join the University of Leuven as a visiting scientist. We left out the fact that I had not received the reprint request card from Edward De Maeyer—the agreed covert message that a position would be available for me in Leuven. I didn't mention the missing card because I was convinced it had simply gotten delayed in the international mail.

As Hans would explain to me only later, in a letter written in the spring of the following year, he had another reason for being concerned about our defection. The Austrian State Treaty, signed by the four Allied powers, was less than a decade old. At the time we arrived in Vienna, in 1964, Austria was well on her way to making an economic recovery, but the scientific base was still reeling from the damage caused by the loss of Jewish professionals after the Anschluss and the ruin inflicted by the war and ensuing Allied—especially Soviet—military occupation. At the time, Moritsch, at age forty-one already a full professor and chairman of the Department of Hygiene and Microbiology at the university medical school, was hoping to continue benefitting professionally from the good relationship with Blaškovič's Institute of Virology in neighboring Bratislava, and he worried that he would be blamed for our defection.

Though I was not aware of these concerns, it turned out that Blaškovič did remain cordial, realizing that the Moritsches had been unaware of our plans to defect until we arrived at their house in Vienna. Sadly, Hans did not have much opportunity to enjoy the continued good relations with Blaškovič and the Bratislava Institute of Virology because, in a shocking and unexpected twist of fate, he died only a year later, in the fall of 1965, as a result of encephalitis—an acute inflammation of the brain—most likely contracted during work in the laboratory with tick-borne encephalitis virus.

Despite their initial shock, the Moritsches were wonderful hosts. We arrived on a Saturday, and—as Hans promised in his anxiety-provoking phone call to Bratislava—we all went to the Vienna State Opera in the evening to hear a splendid performance of Mozart's *Magic Flute*. (No, I did not bring a tuxedo.) Listening to the lovely music, there were moments when I almost forgot the profound change we had made in our lives that day.

On Sunday, the Moritsches took us to see the sights of Vienna. We visited the St. Stephen's Cathedral, the old Imperial Palace, and we walked through the brightly painted streets of the old town. Compared to the gray and drab Bratislava of Socialist days, Vienna was spectacular. We couldn't miss the fact that shop windows were filled with a variety and selection that was unattainable behind the Iron Curtain. This was Marica's first visit to a Western country, and she was captivated.

The next day, Monday, was the day we were supposed to be returning to Bratislava. After arriving in Vienna, Marica wrote a letter to her brother Ivan in New York telling him the big news. We would have phoned him, but we didn't have his number. We had never needed it before, because making a private phone call to America from Czechoslovakia was unthinkable for both political and financial reasons. And when—upon arriving in Vienna—we tried to get his phone number from information, we were told that it was unlisted. Since the letter would take four or five days to reach New York, we sent a telegram too, listing the phone numbers and address of the Moritsches, asking him to call.

As international phone calls were costly even in Austria, Hans suggested that I come to his office to ring Edward De Maeyer in Belgium. I told Edward the news. He said he was happy for us, but I could sense some apprehension in his voice. Was our decision to defect final? he asked, not unlike the Moritsches a couple of days earlier. When I inquired about the prospect of a position in Leuven, he said he had not been able to get De Somer to give him a firm answer, but now that we were in Vienna he would speak to him immediately, and call us back soon. As Monday drew to a close, we had not heard back from either Ivan or Edward.

On Tuesday, Edward called back. He said De Somer was consulting with administrators at the university. I explained that we did not have a lot of time. We did not want to become a burden to the Moritsches and, besides, our legal authorization to stay in Austria had already expired.

By then we realized that we would not be moving out of the Moritsches' house as soon as we had hoped. Edda Moritsch kept reassuring us, explaining that during the war, when her family's residence was bombed, they were taken in by caring strangers, and this was her opportunity to return the kindness.

Even though the Moritsches' house was spacious and we had a comfortable separate guest bedroom, we felt we were abusing their hospitality. But we had no choice. We had no money to pay for a hotel. Besides, we had left without passports. The only documents we possessed were our driver's licenses and a border pass issued by the Czechoslovak authorities granting us permission to stay in Austria through Monday, October 12. The stamp affixed by the Austrian border guards also indicated that the authorized stay was valid for three days. Even with money, we could not have gotten a hotel room without a passport or valid ID card (a driver's license is not considered a valid ID in Europe). Marica never had a Czech passport. I was issued a passport for my earlier foreign trips, but—as decreed by the Communist government in those days—I was not allowed to keep it in my possession, and was required to surrender it to the authorities immediately upon my return from the trip abroad. In addition, before our departure for Vienna, we had to surrender our personal ID cards to the local police in Bratislava.

Edward called on Wednesday. He reported that De Somer was still having some difficulty making formal arrangements for us. Nevertheless, Edward suggested that we visit the Belgian consulate in Vienna, see someone there, and explain that Professor De Somer, director of the Rega Institute at the Leuven University, was securing a position for me at his workplace. That, Edward thought, should be sufficient for us to receive Belgian entry visas.

Anxious to end a precarious situation we rushed to the Belgian consulate where we were referred to a Belgian consular officer. I explained the situation, mentioning the name of Professor De Somer and the promised position. The consular officer asked to see our papers. We showed him the border passes. "What about your passports?" he asked. "We don't have passports," I said. The Belgian officer became very unfriendly. "You are lying," he said to me, raising his voice. "Professor De Somer would never offer a position to someone like you." Very politely I tried to argue with him. "Look," I said, "I am a medical scientist and my wife and I are planning to defect from Czechoslovakia because we want to live in the free world." That apparently was the last straw. "If you don't disappear from here immediately," he said, "I will call the Austrian police and have you deported to Czechoslovakia." We left in a hurry.

―――――

Fortunately this was the low point in our defection journey; the tide started to shift—largely because of help extended us by Edward and Jaqueline De Maeyer. Through an Austrian contact, the De Maeyers first arranged for us to move to a modest but comfortable furnished studio in Vienna. Then, through a friend at the American embassy in Brussels, Edward made an appointment for us with an American official at the US embassy in Vienna.

After our frightful experience at the Belgian embassy, we were apprehensive about following up with the appointment at the US embassy, but the person we met was kind and helpful. The embassy official confirmed what we knew already—that Austria is not the best place for defectors from Communist countries because in 1955, in order to placate the Soviet Union and get the Russian army out of the country, the Austrian government was required to accept a pledge of permanent neutrality. They promised never to join NATO or take sides in international conflicts between the two blocs. Bordering Communist countries in the northeast, southeast, and east, the Austrian authorities were trying hard to avoid arousing the wrath of their Communist neighbors.

When we explained that our goal was to emigrate to the US, the official advised us to cross the border to West Germany, formally called the German Federal Republic, where it would be easier to apply for refugee status than in Austria. Once we received refugee passports in Germany we could apply for US immigration visas, he said.

The embassy official even gave us advice on how to get through the Austrian border to West Germany. He said that in his experience it is the German, not Austrian, customs and security officials that control the border crossing on the main highway from Salzburg to Munich. Most vehicles are simply waved through without being stopped. But, in case we were questioned by the Germans, the US official told us we should explain to them that we were planning to apply for refugee status in Germany.

He also suggested that a good place for us to take up temporary residence in Germany would be Frankfurt because of the proximity of a major US consulate where we would be applying for our visas and have interviews. Finally, he gave us contact information for an American NGO in Frankfurt that might be helpful to us with various formalities and with arranging our travel to the US once we reached that stage.

There were many other people in Vienna who were helpful and kind to us. Even after leaving the Moritsches' house—where we spent six nights instead of the two we had originally planned—Hans arranged that every day a different person from his department would take Marica and me out for lunch in order to help keep our morale up. One colleague in Moritsch's department we had gotten to know well was Christian Kunz, a virologist who, like Moritsch, was working on tick-borne encephalitis virus.

One day, instead of taking us out, Christian invited us for lunch to his parents' house. We were shown to a luxurious apartment furnished with period antique furniture and decorated with exquisite art objects. Marica made an impression by correctly identifying a Gothic Madonna from Bohemia. Lunch was served by a maid in uniform—the first time in our life that we had seen a uniformed servant in a private home. Later we were told that Christian Kunz's family owned large tracts of farmland in the Austrian countryside.

Christian's wife was American. Margaret and Christian had met while he was spending a year as a postdoctoral fellow at the Rockefeller Institute for Medical Research (now Rockefeller University) in New York City. In a weak moment Christian confided to me that he decided to marry Margaret, rather than any of the many Austrian women who would be eager to do so, because Margaret did not know about his family and he could be certain that she was not marrying him for his wealth. At the time we met Margaret, she was still struggling to learn German.

One day Margaret invited us for lunch at a typical Viennese restaurant, Griechenbeisl—located smack in the center of the old town's "first district"—which prided itself on being the oldest inn in Vienna. Over Viennese sauerbraten served with my favorite kind of dumplings, Margaret was telling us what to expect when, one day, we would make it to New York. Marica and I hung on her every word. Some pieces of advice I remember getting from Margaret were "Don't ride the subway after ten p.m.," "Don't walk in the park after dark," and "Avoid walking alone north of Ninety-sixth Street."

Margaret also gave us the name and phone number of her cousin and her husband, Helen and Bill Garrison. She would write to Helen and Bill about us, she said, and we should make sure to contact them when we got to New York. We still count the Garrisons among our close friends.

Margaret Kunz visited us in New York a few weeks after our arrival in America. That was the last time we saw her. Shortly after her visit she was diagnosed with cancer that would soon take her life. As for Christian, I used to run into him during scientific meetings in Vienna and elsewhere. I haven't had contact with him for about a decade, but I read that in

2007, on the occasion of his eightieth birthday, he was honored for his important contributions to the development of a vaccine against tick-borne encephalitis virus.

Having spent two long, turbulent weeks in Vienna, we were ready to move on to Germany, but we had no money to buy gas or food and we had still not heard from Marica's brother Ivan. It took some courage to ask Christian and Margaret for a loan. We did not know when, if ever, we would be able to repay them. They lent us two hundred dollars—a huge amount of money for us in 1964. With this infusion of cash we were able to fill the tank of our little Škoda and take off in the direction of Frankfurt.

Crossing into West Germany, as predicted by the US embassy official, we did not encounter any Austrian border guards, but seeing an unusual-looking car with Bratislava license plates and the "CS" (for Czechoslovakia) marking, a German officer pulled us aside. When he asked for documents, we handed him our expired border passes to Austria. He looked at the passes, then looked back at us and said, "You are not planning to return to Czechoslovakia, are you?" We shook our heads in agreement. He let us pass. We arrived in Frankfurt a few days before Halloween.

———

Our first order of business after arriving in Frankfurt was to register with the German authorities. The American NGO whose contact information we received at the US embassy in Vienna referred us to a German government refugee agency in the nearby town of Oberursel. We filled out lengthy applications, submitted photos, completed interviews, and were then told we would be called when a decision was made.

We were concerned that we had not yet heard from Ivan, but figured that he must be away on one of his exotic vacations. Indeed, when we called Hans Moritsch after arriving in Frankfurt he told us that Ivan had contacted him, anxious to find us and to help. When we finally heard from Ivan, we learned that we had been right, he was on vacation in

Mexico. He immediately assured us that he would lend us money and provide whatever assistance was necessary to secure US immigration visas.

Waiting for the issuance of our German refugee passports seemed to take forever. We needed the passports to even apply for visas to the US, and every day that passed we grew more anxious. With the money we borrowed from Ivan we were able to rent a small, furnished studio and buy food. We tried to keep ourselves occupied by going to museums, but we were not in the right mood for enjoying art. We paid almost daily visits to the university library in order to keep abreast of developments in our fields—and library access was free—but it hardly filled the dreary winter days.

There were periods when our morale sank quite low. We missed our jobs, our friends, and family. We had no idea how long it would take before all our documents were issued and we would be able to leave for the US. Would it be weeks, months, or perhaps even a year or more? There were so many refugees from Eastern European countries who had been idling in refugee camps for years. In general, Marica coped with the situation better than me, and she helped to keep my spirits from sinking too low. Even though we had no money to spend, she enjoyed feasting her eyes on the shop windows and admiring the Christmas decorations in the department store windows. It helped me that she was not sinking into depression.

While in Frankfurt we received an assignment from Ivan. He decided to buy a new car directly from the Mercedes-Benz factory near Stuttgart. Would we be willing to pick up the car on his behalf? And if we were to clock a few hundred miles on the car before having it shipped to America, the US customs would be lower as it would no longer be considered a new car. Besides, Ivan thought, we could use the car to visit some interesting places in Germany. We obliged, of course, even though we were very concerned about a possible accident, given that it was not our car and that we were still living as undocumented aliens, with Czech drivers' licenses our only identification. For a couple of weeks we were penniless refugees driving around in a sleek new Mercedes. And then it was shipped off to Ivan in New York.

I had heard that my friend Bobby Lamberg, who defected in 1957, might be living in Bonn, then the capital of the German Federal Republic. At a long-distance telephone center in Frankfurt, we looked up Robert F. Lamberg in the Bonn telephone directory and, lo and behold, there he was! On a freezing December day, we had a joyful reunion with him and his wife.

Finally, we received word that our passports (marked *Fremdenpass* or "aliens' passport") would be ready for us to pick up on Christmas Eve. We considered the coincidence a good omen. With the passports in hand, we felt more optimistic that before long we might be able to leave for the US. Now we were in a position to apply for US entry visas—except that the American consulate was closed for the holidays until after New Year's Day. The period between Christmas and New Year's was particularly difficult. All of Frankfurt shut down for the holidays, the weather was cold and miserable, and there was nothing for us to do but to sit in our claustrophobic apartment.

We still had our little Škoda automobile. Now that the prospect of securing US visas appeared more realistic, we decided to try to sell the car. It was clear to us that we could not get more than $200 to $300 for it, but even that would have been very helpful. There was only one hitch: it turned out that we couldn't sell the car because it was registered in the name of my mother!

Of course, I should have thought about that before leaving Czechoslovakia, but my family and I had always considered the car to be mine. Now we started to worry that we could be accused of stealing my mother's car, which might even affect our ability to get immigration visas to the US. We asked the NGO that had already helped us for advice, and they contacted the local police to inquire whether they would bring the car to the Czech border and return it to the Czech authorities.

As unlikely as this sounds, the German police agreed to do just that. Even more unexpected was that after some legal wrangling in Czechoslovakia, the car was returned to my mother as the legitimate owner. To prove that she was an authorized owner, my mother produced her valid driver's license. She did not reveal the fact that she had never actually driven the car.

Both Marica and I wrote a lot of letters—Marica mostly to her brother in New York, describing her feelings and new experiences, and I to my professional contacts in the US inquiring about possible job openings.

Soon I started receiving encouraging replies. I heard from Werner Henle, an émigré from Nazi Germany and prominent virus researcher at the Children's Hospital in Philadelphia, who, together with his wife Gertrude Henle, would later identify the causative agent of infectious mononucleosis, a condition colloquially known as the "kissing disease." The Henles had heard about me from their colleague Kurt Paucker, a participant of the Smolenice interferon conference I had organized only a few months earlier, and they indicated their readiness to offer me a research position when I got to the US. I received a similar response from Edwin Kilbourne, another Smolenice conference participant and well-known influenza virus expert at the Cornell University Medical College in New York City.

Another lead for a position came from Marica's brother. At the time, Ivan worked as an anesthesiologist at New York University Hospital. Before a surgical procedure—as Ivan was sharing the story about the defection of his sister and brother-in-law with the surgical team—he learned from the team's head, the prominent neurosurgeon Joseph Ransohoff, that NYU's Department of Microbiology was looking for a virologist.

Ivan relayed the news to me and encouraged me to send my resume to the recently appointed chairman of the Department of Microbiology at NYU, an Australian-born, British-trained bacteriologist, Milton Salton. About a month later I received a reply; Dr. Salton wrote that, provided I could obtain a visa and move to New York within a reasonable time period, NYU School of Medicine was ready to offer me the position of an assistant professor of microbiology with an annual salary of $12,000. That was not only a fortune by my Czechoslovakian criteria, it was a fair offer by American standards. In his letter Dr. Salton also mentioned that NYU would make a contribution toward our relocation expenses.

NYU's offer was attractive for other reasons. It was the only one of the three offers that mentioned a specific position and salary; the other ones were more vague. We—especially Marica—had a preference for settling in New York City, not only because Ivan lived there, but also because of its reputation as the most cosmopolitan city in the US, which we thought would make it easier for us to adjust.

In addition, the offer by NYU was the only one of the three that held out an independent faculty position, which I perceived as a plus, albeit also as a potential risk. Working alongside established scientists such as the Henles or Ed Kilbourne would likely have been easier than setting out on my own at NYU. However, accepting the NYU offer provided the opportunity to establish my own independent laboratory and to build my research program. I decided that NYU was worth the risk. Later, I learned that an important reason Dr. Salton extended such an attractive offer to me was a recommendation he received from Alick Isaacs, the discoverer of interferon. Salton and Isaacs had known one another in London.

It was quite unusual to receive the NYU job offer without ever being interviewed. In more recent years, hiring a new faculty member has become a much more formal affair, with search committees poring over dozens of résumés and recommendation letters, and multiple candidates invited for repeated visits and interviews. But the 1960s were different. This was still the post-Sputnik era, and the federal government was investing significant sums into research in order to fill the perceived gap between research efforts in the Soviet Union and the US. There were not too many candidates to fill certain research positions, and there was a shortage of trained virologists. This is how, within a period of a few months, I went from being a victim of the Cold War to being its beneficiary.

By mid-January 1965 we received invitations to report for interviews at the American consulate and for medical tests. In those days the allocation of immigration visas was still governed by national quotas; there was a ceiling on the number of visas that could be issued for applicants born in any single country. Our good luck was that very few people were able to get out of Czechoslovakia in those days, and so the quota for people born in Czechoslovakia was wide open. It also helped that we both had university degrees, that I had an attractive written job offer, that we had an affidavit of support from Marica's brother, and that we were considered refugees.

One complication was that during Marica's medical examination the physician discovered what he thought was a lump on Marica's thyroid. Fortunately, a follow-up radioactive iodine uptake test ordered by the physician showed no abnormality. Two weeks later we received word that our immigration visas were ready and we could pick them up at the consulate. The earliest flight from Frankfurt to New York we could book was on Pan American Airways on February 4, 1965—a date forever embedded in our memories.

Chapter Ten

America is Different

Monday, February 8, 1965, the fourth day after our arrival in New York City. I was having my first meeting with Dr. Milton R. J. Salton, chairman of the Department of Microbiology, in his office at the NYU School of Medicine on First Avenue near Thirtieth Street in Manhattan. Milton Salton, then in his early forties and already renowned for his research on bacterial cell walls, was an extroverted, convivial person. Australian by birth, he sounded and acted more like a mildly eccentric, old-world, upper-class Englishman, displaying a quintessentially British sense of humor. He smoked cigars incessantly and, as I found out only later, toward the end of his working day he also liked to have a gin and tonic—or two.

We had been having a cordial conversation and I had the feeling that all was going well. Given that he had promised me employment sight unseen, I sensed that Dr. Salton was reassured to find that I was a normal-seeming person and that I was able to communicate in English. Dr. Salton chatted with me about some Czech scientists he had met, and asked me about my experiences after the defection from Czechoslovakia. We talked about my research plans, and I told him I wanted to continue working on interferon.

"May I see my future laboratory?" I asked at one point.

"Let's do that some other day. We didn't have a chance to have the place tidied up for you," replied Milton Salton.

I was disappointed, but did not consider it appropriate to protest. Dr. Salton asked me to come back at the end of the week at which time, he said, we would also discuss more details of my employment.

I came back to see Dr. Salton a few days later, as agreed. By then I was anxious to begin work. Not only did I want to continue my research, I also needed to start earning a living. Since leaving Czechoslovakia almost four months earlier, we had been borrowing money left and right, and by the time we arrived in New York City our debts had reached a few thousand dollars. But Dr. Salton was not in a hurry. He seemed to enjoy conversations about a multitude of subjects, except the specifics of my employment.

Finally, I gathered the courage to ask whether he had decided on my starting date. He had his secretary make an appointment for me with the Personnel Department (the term "human resources" had not yet been invented). Even though the appointment was set for the following week, I was relieved. Then I asked again to see my future laboratory. "We had the space cleaned, but I should warn you that it is still not completely ready," he said.

We walked there together. It turned out that my future laboratory was going to be in a space that, until recently, had housed experimental rats and mice. The cages were gone, but not the smell of animal waste. The space, covered from floor to ceiling with beige tiles, consisted of one fairly large room with a glass-enclosed cubicle and an adjacent smaller room that, Dr. Salton pointed out, I could use as an office. Figuring that the odor would eventually subside, I thought the space was quite adequate.

"Where will the laboratory equipment that I need come from?" I asked. Dr. Salton said that he could find an incubator for me. As for my other needs, I would have to do what everyone else in my situation did in America—write and submit grant applications to secure funds for my research. Once I received the grant money I could buy equipment and supplies, hire a technician, and perhaps recruit graduate students or post-doctoral fellows.

This was a new concept to me. In Czechoslovakia there were short-ages of equipment and laboratory supplies, but everyone received some money for research from the administration and we did not have to write grant applications.

Parenthetically, it is worth pointing out that hiring practices have changed profoundly in academia in the US since the days when I was hired over fifty years ago. On the one hand, the recruiting process for new faculty is now much more formal and stringent, with multiple levels of approval required for each appointment. On the other hand, it is common that newly hired faculty members are offered recruitment packages that may range from something like one million dollars in research support for a junior faculty member to eight-figure sums for the most senior investigators. Other perks, like subsidized housing, are also quite common.

Getting me fully processed for employment took an additional ten days, and I did not get my first paycheck for another five weeks—which was almost two months after our landing at JFK Airport. In the meantime our debts continued mounting, but at least we could see the light at the end of the tunnel.

Meanwhile there were other surprises in store for me. On the day I officially started work, I stopped by Dr. Salton's office. "What are the working hours here?" I inquired. Dr. Salton explained that technicians and administrative employees were generally expected to work from nine to five, but no such rules existed for faculty, graduate students, or post-doctoral fellows. "How much vacation am I entitled to?" "As much as you want," was Dr. Salton's response. I knew that I would be working long hours every day including most weekends and not taking lavish vacations, but still, what a refreshing difference from the Socialist system where everything was tightly regimented!

———

The day we arrived from Frankfurt at JFK Airport was a glorious, sunny winter's day. We moved into Ivan's one-bedroom Manhattan apartment quite close to the NYU Medical Center, where Ivan worked as an attending anesthesiologist. Marica and I slept in the bedroom, Ivan—unmarried at the time—on a sofa in the living room. As much as we wanted to move into a place of our own, we were not able to sign a lease without first completing all of my employment formalities with NYU.

We used the free time before the official start of my job to look for an

apartment. When we arrived in New York, there was a glut of newly constructed apartment buildings in Manhattan and, even though rents were
not cheap by 1965 standards, some buildings offered two months rent-free
to tenants willing to sign a two-year lease. Strapped for cash, we thought
this would help us recover financially. As soon as my employment was
officially confirmed by NYU, we signed a lease on a one-bedroom apartment in a brand-new high-rise building at the corner of Thirty-Second
Street and Second Avenue, a stone's throw from NYU Medical Center.
It helped tremendously that the NYU administration confirmed they
would keep the promise to help with our relocation expenses. The money
for relocation was almost enough to wipe out our debts.

The apartment did not have much character, but it was on the eighteenth
floor, with huge windows, and had an unobstructed, spectacular view of the
Midtown Manhattan skyline; the Empire State and Chrysler buildings com-
peted for our attention. We owned no furniture, and we figured that the
view was the substitute for having nothing to look at inside our apartment.
So, after two weeks at Ivan's, we packed our two suitcases, bought a bed,
sheets, pillows, and towels at the B. Altman & Company department store
and moved into our brand-new apartment. We were ecstatic.

Moving into the new apartment was the easy part. The adjustments
to life in the new country were more complicated. We thought our
English was adequate, but it didn't take us long to discover that we had
trouble understanding people who spoke with regional accents. When
we listened to the weather forecast on the radio we had no idea what "30
degrees Fahrenheit" would feel like. Buying the right amount of ham in
the supermarket created a problem because ounces and pounds meant
nothing to us. And the aggressive sales practices that were common in
this country often took us by surprise. When I called AT&T to order
telephone service, we were impressed that we could have telephone lines
installed the next day, but we ended up with three wall-mounted light-
green Princess telephones, even though one or two standard phones
would have been perfectly sufficient.

And then there were the social interactions. About a week after our
arrival in New York City, Ivan told Marica and me that a friend had

invited him to a cocktail party and asked that he bring us along. We had never been to a cocktail party—a form of entertainment that did not exist in Bratislava. The party was hosted by a friend of Ivan's who was married to a well-to-do husband, in their apartment on Central Park West.

When we arrived there were about fifty people standing in the living room in groups of three or four, all having animated conversations. We were introduced to some of the people in the room. Everyone was cordial, but we had trouble maintaining a conversation, partly because in the noisy room we had a hard time understanding people, but also because we had difficulty with the small talk that seemed required.

We were offered martinis (at the time, wine was still not commonly served at parties in America), a drink that was new to us, and that, at first, we did not find pleasing. A waiter was serving shrimp from a tray. We had never seen let alone tasted shrimp, and we did not find the unfamiliar smell inviting. Overwhelmed by the new experiences, Marica and I retreated to a corner near the window from where we could quietly admire the view of the Central Park reservoir. Our hostess approached us with a big smile. "Are you having fun?" she asked.

On the first day at work, Dr. Salton personally introduced me to everyone in the Department of Microbiology, which at the time consisted of about ten faculty members and some fifty other people, including graduate students, postdoctoral fellows, lab technicians, and administrative employees. They were all extremely welcoming. As promised, Dr. Salton had had an old incubator moved into my empty laboratory, but it was of a type that could not be used for growing tissue cultures, and so was of little use to me. The smaller room—my office—was now equipped with an old metal writing desk and a used desk chair. A reassuring sign of progress was that the foul smell was almost gone.

I knew that the first order of business was to work on finding grants. Several of my colleagues in microbiology offered to help familiarize me with the business of preparing grant applications.

At the same time, I was in the midst of an extraordinary group of scientists I wanted to get to know. NYU School of Medicine in those days was (and still is) a remarkable place. Chairing the Department of Biochemistry was Severo Ochoa, a refugee from Franco's Spain, who had earned the Nobel Prize for Physiology or Medicine for his contributions to the elucidation of the mechanisms of RNA and DNA synthesis, and for deciphering the genetic code.

Especially strong was the field of immunology, represented by, among others, three towering figures: Baruj Benacerraf, born in Venezuela to a wealthy banking family of Sephardic Jews from North Africa, who would go on to win the Nobel Prize for Physiology or Medicine in 1980. He was awarded the Nobel for the discovery of immune response genes—genes that control the immune system's ability to respond to infections or other foreign material. Edward Franklin, another prominent immunologist at NYU, received the honor of having a disease named after him, but, sadly, would die of a brain tumor in 1982 at age fifty-four.

And then there was Michael Heidelberger, already in his late seventies when I joined NYU, who is considered the father of modern immunology. He is credited with many basic discoveries, including demonstrating that antibodies are proteins and that complex sugars in microbes can act as antigens, meaning they can stimulate the production of protective antibodies. Working full-time until close to his death at age 103, Michael was also a colorful person and a great raconteur, always at the ready to come up with a new story from his long and productive life. Michael earned every conceivable scientific honor—including the Lasker Award and the National Medal of Science—with the exception of the Nobel Prize. In one of his stories, he related to me why he thought he had been passed over by the Nobel Prize Committee—he had once turned down an invitation to give a lecture at the Karolinska Institute in Stockholm. (Selections of Nobel Prize winners in Physiology or Medicine are made by professors of the Karolinska Institute.)

Michael's joie de vivre at his advanced age was admirable. When he was about ninety-four, he was told that he needed to have a heart valve replaced, which required open-heart surgery. The heart surgeon told him there were two options, he could use a pig heart valve or an artificial one, which one would he prefer? "I don't care," said Michael, "but give me the valve that will last longer."

Another renowned faculty member, until his departure in 1969, was Lewis Thomas, a noted experimental pathologist as well as author of *The Lives of a Cell: Notes of a Biology Watcher* and of many other award-winning books of essays. At NYU, Lew Thomas served successively as chair of the Department of Pathology, chair of the Department of Medicine, and dean of NYU School of Medicine. He was a great thinker and humanist. Encounters with him were always stimulating.

———

One of the people who played a role in bringing me to NYU was Joseph Ransohoff, chair of the Department of Neurosurgery. After hearing about me in the operating room from Ivan, Joe had mentioned my name to Dr. Salton and the then-acting dean, Saul Farber, which led to the job offer. Joe was a world-class neurosurgeon, an almost legendary figure— he inspired the medical drama series *Ben Casey* that ran on ABC television from 1961 to 1966. He had not only pioneered surgical treatments of brain tumors, he was also interested in learning about their causes.

When I first met Joe, a few days after starting at my new job, he confided that he was hoping I would come up with ways to identify viruses responsible for some of the deadly brain tumors he was interested in, such as glioblastoma. At the time it was suspected that viruses might be responsible for some human cancers, because certain tumors in animals were known to have a viral etiology. But there was still no specific virus known to be implicated in any type of human cancer. To motivate me toward expanding my interests into the field of experimental cancer research and the role of viruses in cancer, Joe introduced me to two well-known cancer researchers, Lloyd Old and Edward Boyse, at the Memorial Sloan Kettering Cancer Center. Lloyd turned out to be invaluable at later stages of my career when I became interested in the tumor necrosis factor protein, or TNF.

Unfortunately, I never was of much help to Joe's quest to identify viruses as causative agents of brain tumors—this was not a project I thought I could successfully tackle. To this day, fifty years later, there is still no virus positively identified as the cause of brain tumors in humans. However, progress has been made in identifying viruses that cause some other types of human cancer, for example Epstein-Barr virus (the causative agent of infectious mononucleosis) is also known to be the cause of Burkitt's lymphoma—a cancer occurring in children in equatorial Africa; hepatitis B and hepatitis C viruses are the cause of primary liver cancer; and human papilloma viruses have been identified as the cause of cervical cancer of the uterus. Altogether, it is now estimated that about 20 percent of all human cancers have a proven viral etiology and it is likely that more types of human cancer will turn out to be elicited by virus infection. In his general instincts, Joe Ransohoff was on the right track.

In my first weeks, I also met scientists outside NYU, including two well-known researchers at the Albert Einstein College of Medicine in the Bronx. One was Wolfgang "Bill" Joklik, a prominent virologist who was interested in interferon research. Using Bill as a sounding board for some ideas I had for my future interferon research helped to clarify my thoughts about what I would include in the grant application I was about to start writing.

Another scientist I met at Albert Einstein was Barry Bloom, then still a young microbiologist and immunologist, soon to become a widely recognized expert in infectious diseases, vaccines, and global health. Though not working on interferon, Barry was interested in my work because of his involvement with another protein important in host defenses against microbes, called macrophage migration inhibitory factor.

My first grant application took about two months to prepare, not very long considering the fact that I was new to this business. It was already known that interferon was a cell-made protein—that is, a protein produced and secreted from cells—but the production happened only after the cells' infection with a virus. In the application, I proposed to investigate what made cells switch on interferon production. It was suspected that cells must have one or more genes for interferon, and that the expression of this gene or genes was somehow turned off until a virus infected a cell. However, why and how a virus would turn on the generation of interferon was not known, and I was proposing to find some of the answers.

The answers were important because at the time interferon was the only known substance that selectively inhibited the multiplication of viruses without harming cells. There was already some support for the idea that interferon was important in the body's defense against viral infections. I believed—as did others in the field—that as we learned more about how interferon was made and how it worked, we might be able to use interferon for the treatment of viral infections. In that first grant application, I wrote, perhaps a little too optimistically, that interferon might turn out to be as important for the treatment of viral infections as penicillin along with many other antibiotics are for the control of bacterial infections. It is now known that interferon is important in the body's resistance to many virus infections, but it is used relatively infrequently as a therapeutic agent for viral diseases.

I knew in writing my application that not everyone would be persuaded about the importance of the work I was proposing to do. The interferon molecule had not yet been isolated in pure form, and consequently there was no physicochemical proof of interferon's existence. All the available evidence rested on indirect biological assays. Some scientists were not convinced, and detractors commented that a better name for interferon might be "imaginon." In addition, the research methods available for the work I was proposing were still quite crude, again depending more on indirect evidence than precise molecular analysis.

Heeding the advice of Dr. Salton and others to apply to multiple agencies and hope to have my choice of grant offers, I submitted the application to the National Science Foundation (NSF) and the New York City Health Research Council, in addition to the National Institutes of Health, the major federal funding agency of biomedical research. Responses, I was told, could take about six months. I was hopeful but not very confident that my grant application would be funded. In the meantime I was getting ready to start teaching, attending lectures, and doing as much reading in my field as possible.

As I was settling into my new life as a research scientist and medical educator in America, some colleagues and acquaintances asked whether I had considered the practice of medicine as a career option instead of devoting myself to laboratory research and teaching.

I remember vividly a conversation with Ernst Wertheimer, a friend of my parents who with his wife, Lilly—my mother's best friend in Bratislava—had emigrated to Israel but later moved to the Boston area. Ernst, a gynecologist by training, was in his fifties when they moved to the US to join their daughter and grandchildren. Perhaps because of his age and because he never completely mastered the English language, Ernst had failed to pass the examination all foreign medical graduates need to get through in order to be allowed to work with patients. Unable to practice medicine, Ernst had to supplement his small Israeli pension by working as a technician in a hospital lab in Brookline, Massachusetts. Lilly, still attractive and lively, worked as a receptionist in a restaurant.

Ernst and Lilly came to visit us when they learned from my parents that we had settled in New York City. As close family friends, they didn't hesitate to ask personal questions.

"How much do you make?" Ernst asked bluntly. I answered truthfully. "That's not bad," said Ernst, "but do you realize that you could make several times that amount in medical practice? You are young, your English is good, I am sure you would have no difficulty passing the ECFMG examination [the exam Ernst had failed]." He went on to point out that in science the success of my career would always depend on my ability to secure grant support. "You will be working your ass off, and you will never make enough money to live a really comfortable life." (I should mention that in the 1960s doctors enjoyed a very high social standing—probably more so than today—and, in relative terms, their average earnings too were higher than in the days following the introduction of managed care.)

I knew that he had our best interests in mind. But, after thanking him, I told Ernst that I had made up my mind. I knew there was no guarantee that I would succeed as a research scientist, but I was willing to take the risk. "I probably could become a decent doctor and earn more money," I said, "but I love what I am doing and I don't mind if I never get rich."

———

At the same time, Marica was trying to find a professional home of her own. There were—and still are—fewer job opportunities in art history than in biomedical science. Nor did Marica have the type of professional contacts in the US that I had been able to establish while working at the Institute of Virology in Bratislava. She sent out letters to museums and libraries in New York City, inquiring about possible openings. Today, most organizations don't bother to acknowledge unsolicited job inquiries; at best they respond with a form letter. In 1965, people were more courteous. She received many replies, and even though at first no one invited her for an interview, some of the letters offered words of encouragement and assurances that if a suitable opening should arise, she would be notified. The director of the Frick Collection went as far as to suggest, in a personal letter, that Marica's chances would improve if she had her CV revised and reformatted so that it conformed to American standards. She did!

During our stay in Vienna, Margaret Kunz—the American wife of my Austrian colleague Christian—had given us the phone number of her cousin in New York City. We contacted Helen Garrison and her husband Bill shortly after our arrival. Helen was very eager to help Marica in her job search—and in the process of adjustment to life in America in general. When we met her, Helen was at home with her young twin daughters, but earlier she had worked at the Brooklyn Museum. When it became clear that finding a paid position for Marica might take some time, Helen asked whether Marica would be interested in working temporarily as a volunteer at the Brooklyn Museum, explaining that such experience might be helpful in Marica's search of a permanent job. Marica agreed enthusiastically and by springtime, thanks to Helen's recommendation, Marica started working as a volunteer in the Brooklyn Museum's library.

Marica's spirits improved. She too would now come home in the evening with new experiences to share with me over the dinner table. By then we did in fact own a dining table—like the few other pieces of furniture we had in the apartment at the time, it had been acquired from a thrift shop run by the Salvation Army.

Then, before the summer ended, Marica secured a permanent job. In fact, she suddenly received two job offers. First she had been invited to an interview at the New York Public Library—one of the places where she had submitted her improved résumé—and shortly thereafter she was told that she could have a position there. She would have said yes, had she not also been told of a possible opening at the Metropolitan Museum of Art. Some months earlier, Helen Garrison had referred Marica to a curator at the Met who had talked to Marica and then guided her to personnel, where she filled out a job application for the position of a cataloger, and where she was told she would be contacted if an opening arose.

It turned out that the position available at the Met was not for a cataloger, but for a clerk-typist at the museum library. Marica was offered the job. Museums were not and still are not known for paying generous wages; the gross annual pay in the position offered to Marica was $3,750 per year, or about two dollars per hour. This was a meager compensation even in 1965 when the federal minimum wage was $1.25 and a gallon of regular gas cost less than thirty cents. Despite the low pay and unattractive job title, Marica accepted the offer, hoping that if she did well she would eventually move to a professional position. Besides, even though Marica's salary was low, it helped significantly with our expenses and the remaining unpaid debts. To this day, one piece of advice Marica often gives to people seeking their first jobs is that in the fluid American job market it is important to get started in some position, even if that position is not the most desirable one.

Marica spent about seven months working in the Met museum library before being promoted to the professional position of cataloger, more in line with her background and expectations. Fortunately, her position of clerk-typist did not require a great deal of typing. The head of the library at the time, Elizabeth Usher, recognized that Marica's familiarity with several Slavic languages in addition to German, French, and Hungarian was more of an asset than her not-so-stellar typing skills.

Marica quickly became popular at the library, partly because of a small faux pas committed on one of her first days on the job. She answered a telephone call, which, it turned out, was for an older unmarried female colleague. In continental Europe a professional woman of a certain age would never be addressed as "Miss," no matter her marital status. "Mrs. Wozar is on a lunch break," Marica announced. The next day, Miss Wozar ("Ms." was not yet in use in 1965) received a congratulatory call on the occasion of her marriage.

The fall marked for me the beginning of a new experience—I was asked to help with the teaching of the laboratory section of the microbiology course for medical students. While I had worked as teaching assistant in microbiology during my medical school days, the standard of teaching at NYU was higher. In those days microbiology was taught one full semester in the second year of medical school, with two two-hour lab sessions scheduled per week. The students learned the basics of diagnostic micro-biology and, unlike in Czechoslovakia, they also performed some fairly sophisticated experiments that taught them principles of the science of molecular microbiology.

I was taken under the wing of Lane Barksdale, a kind if somewhat eccentric senior professor in the Microbiology Department. Every year, when he gave a lecture on yeast and fungal infections, Lane would bring a wicker basket filled with mushrooms (also a kind of fungi, of course) that he would throw into the auditorium during the lecture. Lane loved teaching, even though some students complained that his lectures were incomprehensible.

Toward the end of the year I received the first reply from one of the agencies where I had submitted my grant application; it was from the New York City Health Research Council, an organization that ceased to exist when the financial crisis hit the city in the 1970s. I was thrilled when I learned that they were willing to fund my research. More good news was on the way: the NIH and the National Science Foundation, too, were ready to give me the research support I had applied for. I don't know if it was compassion for a refugee from behind the Iron Curtain, or that they believed in the importance of interferon, despite the detractors. I accepted the grant support from the NIH and—because it would have been improper to accept funds for the same project from more than one source—I declined the other two grants, though with gratitude for their offers.

Eager to get started with my research project, I placed orders for several pieces of equipment. I did not have to purchase a carbon dioxide (CO_2) incubator—an expensive piece of equipment needed for tissue culture work—because a few weeks before I had received word that my grant was funded, I was approached by a young NYU colleague who introduced himself and said, literally, "I have a CO_2 incubator, can I move it into your laboratory?" At the time I was all by myself in the lab, and without much hesitation I invited the colleague, Alvin Friedman-Kien, to join me. (More about Alvin very shortly.) The funds provided by the NIH grant were sufficient for only one salaried position, which I used to hire a technician. I found a competent young woman, Millie Varacalli, an immigrant from Puerto Rico who then worked with me for more than a decade.

A few weeks later, I was approached by a newly admitted PhD candidate who inquired if I would consider becoming his thesis adviser. The student was Mun H. Ng, a native of Hong Kong who came to the United States planning to earn his PhD degree in microbiology, and then return home. Interested in working in the field of virology, Mun asked Dr. Salton who in the Department might be available to become his adviser, and Salton told him that I was the only person researching viruses. Soon another graduate student, Toby Rossman, knocked on my door to inquire about the possibility of joining my lab. Toby had completed most of her required coursework and was ready to start the laboratory research necessary for writing and defending her PhD thesis. She had read some of my publications and was interested in the work I was doing with viruses and interferon.

I was quite stunned that any self-respecting student would consider joining a lab that was still far from being fully functional, but sensing that both Mun and Toby were smart and personable, I agreed to give them (and myself) a chance. A third student who joined the lab in the summer of 1966 was Douglas Lowy. Doug had stood out as a bright and inquisitive medical student in the laboratory section of the microbiology course I had been teaching. Remembering how I had become interested in research as a medical student in Bratislava, I asked Doug if he might consider joining my lab for a research internship during the summer, which he readily agreed to do.

And then there was Alvin Friedman-Kien, who had talked his way into my lab by offering to share his CO_2 incubator with me. Alvin was an MD completing his fellowship training in dermatology. Earlier, he had spent time at the dermatology branch of the NIH in Bethesda, Maryland, and during that period he worked in the laboratory of Wallace "Wally" Rowe, a highly regarded virus researcher with a number of discoveries to his credit. Restless and ambitious, Alvin not only aspired to become a leading clinical researcher-dermatologist, he was also a passionate collector of art and antiques from many different cultures and periods. Moreover, he was successfully investing in real estate in New York City and occasionally making a profit by buying and selling art and antiques. Alvin became indefatigable in his efforts to help me—and later Marica too—to get adjusted to life in New York and to introduce us to the art and culture of this country.

One day during our lunch break Alvin offered to show me the Seagram Building designed by the German-born architect Ludwig Mies van der Rohe. Alvin took me inside the building into the fabled Four Seasons restaurant—not to eat there, of course, but to show me the interior, conceived by influential American architect Philip Johnson, now a designated New York City landmark. Alvin pointed out the sculpture by Richard Lippold (then a friend of Alvin's) that still hangs from the ceiling above the bar in the Grill Room and the large stage curtain painted by Pablo Picasso adorning a wall near the entrance to the Pool Room. (In 2014 the fragile curtain was taken down for a badly needed cleaning and restoration. It has since been moved to the New-York Historical Society for permanent display.)

Before leaving, Alvin also showed me the richly decorated marble men's room of the restaurant. But he wouldn't let me linger inside the men's room. "Let's get out fast," he said, "so that we don't have to pay them a quarter." Alvin and I have remained close friends.

The small coterie crowded inside the tiled walls of my lab got along splendidly. Mun and Toby went on to successfully complete their PhD thesis research and to publish good scientific papers. Doug came back to work in the lab the following summer and he too had his name included on a publication. American-born Toby and Doug helped me tremendously to gain an understanding of how young people think in this country. All of the people inside my small laboratory were smart and creative and they deserve a great deal of credit for getting my research successfully underway. I can only hope my students—all of them, not just Toby, Mun, and Doug—have learned as much from me as I have learned from them.

In addition to setting up the lab and getting started with the experimental work, I took on another project. Shortly after my move to NYU, I received an invitation to write a monograph on interferon from the European science publisher Springer-Verlag. This was a tempting and flattering invitation because the resulting product would become the very first comprehensive text devoted to the review of the entire field of interferon research. I responded that I was interested but—in view of the fact that I had to complete my grant applications and get my newly established laboratory going—I could not start working on the assignment until the middle of the 1967 calendar year and—if all went well—have it completed a year later. The publisher accepted. I worked in the laboratory during the day and at home, after dinner, I wrote the text of my book longhand. Doug, an art history major in college and a skilled wordsmith, acted as my unpaid editor, helping tremendously to make the 140-page hardcover book readable when it finally saw the light of day in 1969.

Every member of my small laboratory in the mid-1960s went on to build a distinguished professional career. Mun Ng returned to Hong

Kong, as planned, where he eventually became professor and chairman of the Microbiology Department of the Hong Kong University School of Medicine. Toby Rossman went on to receive postdoctoral training and then returned to NYU to become professor of environmental medicine at the School of Medicine.

Doug Lowy completed training in dermatology, then moved to the NIH where he later joined the National Cancer Institute and built a stellar scientific career. Doug made a major contribution to the development of the papilloma virus vaccine—a vaccine widely used to prevent infection with papilloma viruses in preteen girls—and lately boys too—and impart protection against cancer of the uterine cervix and some other malignant tumors caused by papilloma virus infection. In 2014 President Obama awarded Doug and his colleague the National Medal of Technology and Innovation, one of the two highest honors bestowed by the United States government for achievements in science, technology, and innovation. Doug is now acting director of the National Cancer Institute.

Alvin Friedman-Kien went on to establish his own research program at NYU, becoming the first physician-scientist in the early 1980s to identify cases of Kaposi's sarcoma in young homosexual men—a telltale sign of the emerging AIDS epidemic—thus becoming a living legend in the history of AIDS research.

Witnessing the growing success of many of my former students and colleagues has been one of my life's greatest satisfactions.

PART FIVE | New Eyes

Chapter Eleven

Dreaming in English with an Accent

Emigrating from a Communist country without official permission was considered a crime. In due course, Marica and I were tried in absentia in a court in Bratislava and convicted of leaving Socialist Czechoslovakia without legal authorization—well, that was true. We were both found guilty as charged and sentenced to a loss of all of our property and to prison terms—Marica for three years and I for five.

The loss of property was largely symbolic, because we had no savings and we owned nothing of value. That said, the authorities sent marshals to search Marica's father's house, where we had lived, and they confiscated some pieces of furniture we had used, even though they were mostly hand-me-downs from my parents and Marica's father. Before the marshals arrived, my father-in-law succeeded in hiding a few Czech and Slovak modernist paintings and works on paper that Marica had collected. The objects had very little monetary value, but Marica was happy to be reunited with them some years later.

The jail sentences, too, were symbolic because we were not about to return to Communist Czechoslovakia to serve them. When we learned about the sentences, I teased Marica that she deserved the longer prison term, not me, because she was the one who had persuaded me to leave the country. In 1968, during the period of the Prague Spring, all past sentences meted out for "illegal emigration" were rescinded. And in 1991, after the Velvet Revolution that ended nearly forty-two years of Communist rule in Czechoslovakia, the new democratically elected government abolished the law that had criminalized emigration.

Symbolic as they were, the court sentences served as a warning to our friends and colleagues that having contact with us was potentially dangerous. But not even the Communist bosses could expect that our parents would break all relations with us. Besides, our parents were either retired or very close to retirement and not much could happen to them. Yet even our parents were still careful about what they wrote to us because all foreign mail was covertly monitored. Marica's younger brother, who was trying to get admitted to a university, had to be especially cautious, because writing to us could have endangered his future.

For our friends and former colleagues having contact with us could lead to an "invitation" from the ŠtB—the secret police. There were a few friends who still had the courage to write to us—they too had to carefully choose the words they would put on paper—but most others considered it wiser not to risk possible repercussions and decided to forego contact with us. We understood.

For a brief period it appeared that our isolation from the old country might ease. In January 1968—a little over three years after our departure—reform-oriented Alexander Dubček became first secretary of the Communist Party of Czechoslovakia and ushered in a period of liberalization known as the Prague Spring. Initially it appeared that the country might succeed in introducing "socialism with a human face," as the reformists had put it, perhaps a kind of social democratic system without restrictions on freedom of speech, the press, and travel.

Not surprisingly, such reforms were strongly opposed by the Soviet leadership and on August 21, invoking the "Brezhnev doctrine," the Soviet army, with support from the armies of other Eastern Bloc countries, invaded Czechoslovakia, removed Dubček from power, and installed the much more orthodox and Soviet-friendly Gustáv Husák as first secretary of the Communist Party of Czechoslovakia.

A period the Communists dubbed "normalization" ensued, immortalized by the Czech writer Milan Kundera in his novel *The Unbearable Lightness of Being*, which was later adapted into a successful movie starring Juliette Binoche and Daniel Day-Lewis. Oscillating between malaise and agony, it would take another twenty-one years for the Communist system in Czechoslovakia and other Eastern European countries to collapse under its own weight when the Cold War ended in 1989.

Even at a safe distance in New York, watching these events unfold was heart-wrenching for Marica and me. In the immediate and chaotic aftermath of the Soviet invasion and occupation, travel restrictions were suspended in Czechoslovakia and border crossings to Austria and West Germany remained open for several months, enabling an estimated 250,000 Czech and Slovak citizens—roughly 1.7 percent of the country's total population, most of them educated—to flee.

Many of our friends used the opportunity to emigrate and some of them eventually settled in the US. We felt that we had an obligation toward newcomers who, like us some three years earlier, came with few possessions and needed temporary shelter. For a period, our one-bedroom apartment in New York served as a transient home to many individuals, couples, and even small families.

We realized that had we not defected four years earlier, we would almost certainly have been fleeing during the large exodus following the collapse of the Prague Spring, but we would have been four years older and getting our lives started anew would have been more difficult.

We could not go back to visit Czechoslovakia for many years. I returned for the first time nine years after our defection, by then traveling with an American passport. Marica did not go back until some years after that.

———

February 4—the date of our arrival to New York City—has become an important annual fete for us. By the time we had celebrated our third anniversary as adoptive New Yorkers, we felt that we were adjusting. We had less trouble understanding people speaking with regional accents and we started to learn about the ins and outs of national and local politics. We were no longer completely lost when the conversation turned to squabbles between Democrats and Republicans or upcoming mayoral elections. We knew that 75 degrees Fahrenheit was agreeable and 95 was not. We were even getting used to cocktail parties and learned to like shrimp, lobster, oysters, and other types of seafood that had seemed so unappetizing three years earlier. There were, and still are, limits to our

Americanization—to this day we don't have the faintest idea what is going on in a baseball game.

One day, in our third or fourth year in New York, Marica and I made a radical decision. From that day on we would no longer speak to each other in Slovak. We were going to converse only in English. We made this decision because—not unlike other immigrants—in our conversations we started mixing Slovak and English words, creating our own version of "Slovlish," and we disliked that. We wanted to improve our English vocabularies, to learn to communicate on a more sophisticated level in our adoptive language, and we felt that we could achieve these goals more effectively if we switched completely to English.

The first weeks of the implementation of our English-only policy were difficult—it seemed so unnatural for the two of us to communicate in English that we would frequently slip back into Slovak. Over time we got used to it. I even forced myself to do sums in English, and to my surprise, after a while it became perfectly natural. Then one day I realized that I was thinking in English, and when I spoke in Slovak to someone I would mentally translate from English into Slovak—not the other way around—and come up with sentence structures that sounded odd in my native tongue. After a few more years, I realized that I was dreaming in English—with a Central European accent, of course!

We did not make these efforts because we wanted to conceal our European origins or appear to be more American than Benjamin Franklin, but because we did not want to feel like perpetual strangers in our new home. We ran into émigrés from Czechoslovakia, most of them exiles who had left after the Communist coup in 1948. They were cultured and perfectly nice people, but after all these years in the US they were still living in the Czechoslovakian past. Most conversations were about the days back in "our country"; for them Czechs and Slovaks were "us" and Americans were "them." They still hoped that one day after the fall of Communism they would be able to return home, where they would regain their lost property and former positions.

We did not see ourselves as exiles, nor did we envisage moving back to Czechoslovakia even if one day the Communist system were to collapse, which, in any case, we knew would not be happening soon. Fortunately, it was easy for us to get used to feeling at home within the multicultural and multiethnic community that is New York City.

We became US citizens in the spring of 1970 and immediately applied for American passports. Getting citizenship and US passports was important for us. We had received our green cards shortly after arriving to New York, which was perfectly sufficient for living in the US, but travel to most foreign countries was a problem. We were considered stateless and had no valid travel documents other than the German refugee passports that were good for getting our US immigration visas, but not for any other travel. Before becoming a US citizen I was invited to participate in a scientific meeting in London, which turned out to be quite complicated. I had to apply for a reentry permit from the US Immigration and Naturalization Service—a document that looked like a passport, but was cumbersome to use. Even though I had secured a valid British entry visa, I was nearly turned away by the immigration official at Heathrow Airport.

Now, with brand-new US passports in hand, in the summer of 1970 Marica and I made a trip to Italy, with stopovers in Dubrovnik and Vienna. We had a great time. On the way back, as we were about to land at JFK, we reminisced about our flight from Frankfurt to New York some five years earlier. Then we had been excited about the prospect of starting our life anew in an unknown city. This time we were returning home.

I recently read a *New York Times* opinion page article by Costica Bradatan, an émigré from Romania, about uprooting and readjustment, that I consider astute. He wrote,

> When your "old world" has vanished you are suddenly given the chance to experience another. At the very moment when you lose everything, you gain something else: new eyes. Indeed, what you eventually get is not just a "new world," but something philosophically more consequential: the insight that the world does not simply exist, but it is something you can dismantle and piece together again, something you can play with, construct, reconstruct and deconstruct.

Funds received from the NIH enabled me to equip my newly established laboratory and begin studies proposed in my grant application. With my first graduate students Toby Rossman and Mun Ng, we initiated a line of studies that analyzed how interferon synthesis is regulated in cells called "fibroblasts," a cell type common in connective tissues, such as the tissues located under the skin.

In one type of experiment, we used appropriate treatments to stimulate cells to start producing interferon and then added substances that inhibited either cellular RNA synthesis or protein synthesis. (Interferon is a protein but a type of RNA called messenger RNA is required for the synthesis of all proteins.) We were surprised to find that when the inhibitors were added to cells at certain times, they not only failed to suppress but actually enhanced interferon production. Initially we did not know how to interpret these findings, but eventually we came up with the hypothesis that there are proteins in the cell capable of suppressing interferon production, and treatments that dampen the production of these inhibitory proteins more than the synthesis of interferon itself would result in a net increase in interferon production. It would take years of work by us and other groups to confirm that this hypothesis was indeed correct.

In the meantime, we realized that certain combinations of treatments with inhibitors of RNA and protein synthesis, added to fibroblasts that were stimulated to produce interferon with a type of synthetic RNA, resulted in the production of large amounts of interferon—much more than could be generated by infection with viruses or by stimulation with synthetic RNA alone.

Being able to produce large amounts of interferon was desirable for two reasons. First, at the time we were performing these experiments in the late 1960s and early 1970s the interferon protein was still not isolated, purified, or biochemically characterized, and in order to succeed in these tasks it was necessary to find ways to generate as much interferon as possible. Second, we wanted to find out if interferon could be successfully used for the prevention and treatment of viral infections, but there were no methods available to produce sufficient amounts of interferon for use in clinical trials.

When Toby Rossman and Mun Ng left the laboratory after defending their PhD dissertations, a new postdoctoral fellow joined my laboratory. Edward Havell was born to a Czech American family in Berwyn, Illinois, a suburb of Chicago settled by large numbers of Czechs in the early years of the twentieth century.

Ed took over the project aimed at devising methods to improve the production of human interferon in laboratory-grown fibroblast cells while also trying to learn more about how interferon production is regulated. Progress was steady, but quite slow. By then there were two other laboratories pursuing goals similar to ours—one at the University of Pittsburgh and one at the University of Leuven in Belgium. Like us, they used treatments with inhibitors of RNA and protein synthesis along with other tricks in the quest to parlay cells into producing significant quantities of what was called "human fibroblast interferon"—so named because its source was human fibroblasts.

Ed Havell and I invested a great deal of work in perfecting the production of human fibroblast interferon. Part of the effort involved finding the optimal combination of treatments and conditions. Other work focused on finding a cell line that would yield the highest levels of interferon. Since one of the goals was to use interferon in clinical trials, we preferred to use normal cells rather than cancerous cells, even though the latter are usually easier to grow and propagate. The use of cancerous cells for the production of interferon destined for clinical trials could have generated problems with regulatory agencies, because contamination of the final product with DNA or some other molecule from these cells might have posed a danger of transferring cancer to the patients receiving the treatment.

Our preferred cell type were normal human fibroblasts derived from foreskins removed during the circumcision of newborn baby boys, which we procured from the maternity ward of NYU's Tisch Hospital. The initial amount of tissue was very small, but cells isolated from the tiny foreskins could be grown for many generations, eventually yielding billions of cells.

One line of foreskin cells established in our laboratory, shown to be able to generate high yields of interferon, was named FS-4, with "4" denoting the serial number of the foreskin received in our laboratory. Over the years we received dozens of requests from investigators from around the globe to supply them with FS-4 cells so that they could replicate our data and produce interferon in their own laboratories. We complied with every request. Our cells and our methods were instrumental in making possible the generation of human fibroblast interferon for a number of clinical studies and for the complete purification and characterization of the interferon protein.

The methods of fibroblast interferon production we helped to develop in my laboratory also enabled the cloning of a form of DNA called "complementary DNA" (meaning a DNA whose sequence matches that of the messenger RNA) and its adaptation for the artificial production of human fibroblast interferon by recombinant DNA technology. Interferon production by recombinant DNA technology is now the only method used for the preparation of interferon for medical applications. (More about that later.)

Today fibroblast interferon (now known as IFN-beta) is widely used for the treatment of multiple sclerosis, MS for short—a chronic, usually progressive neurological disorder in which myelin sheaths, the protective covers of nerve cells, are severely damaged by inflammation. Though not a cure, the administration of fibroblast interferon reduces the flare-ups of MS and slows its progression.

Fibroblast interferon was first used in MS in the 1970s by Lawrence Jacobs, a neurologist in Buffalo, New York. Jacobs tried interferon because it was (and still is) thought that MS might be triggered by viral infection, and interferon is known to suppress viral infection. However, by the time disease symptoms of MS are present, the damage that might be caused by any triggering virus is fully established and it seemed unlikely that suppressing viral infection at that stage would be beneficial. For these and

Marica's brother Ivan Gerhath who was living in New York City when we moved there in 1965.

With Marica on our first visit to Central Park in New York City in February 1965.

Marica c. 1966 at the Metropolitan Museum of Art, her workplace for thirty-two years.

With my parents in Italy c. 1970.

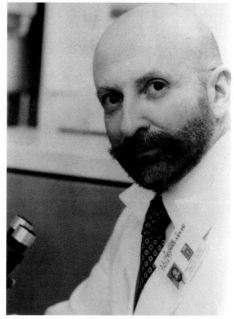

Alvin Friedman-Kien, AIDS
researcher, art collector, and
friend c. 1985.

My colleagues, Edward Havell (second from left) and Mun H. Ng (third from left) in conversation with interferon scientists William E. Stewart II (far left) and Pravinkumar Sehgal (far right) c. 1972.

With colleagues in my laboratory at NYU c. 1988. L. to r.: seated: Yihong Zhang, Luiz Reis. Standing: Tae Ho Lee, Hans-Georg Wisniewski, Jedd Wolchok, Jan Vilcek, Jumning (Jimmy) Le, Jian-Xin Lin, Gene Lee.

With colleagues in my laboratory at NYU c. 1993. L. to r.: Ryutaro Kamijo, John Gerecitano, Paul Schwenger, Deborah Shapiro, Hans-Georg Wisniewski, Jan Vilcek, Adam Goodman, Lidija Klampfer, Gene Lee, Ilene Totillo, Peter Sciavolino, Angel (Phil) Feliciano.

Harold Varmus (left), co-winner of the 1989 Nobel Prize in Physiology or Medicine, with Joan Massagué, winner of the 2006 Vilcek Prize in biomedical science, at the 2006 Vilcek Foundation Awards Dinner.

Diane Sawyer with her husband Mike Nichols, recipien of the Vilcek Prize in filmmaking, at the 2009 Vilcek Foundation Awards Dinner.

L. to r.: Ruslan Medzhitov, Vilcek Prize winner in biomedical science; Tada Taniguchi, my Japanese colleague and friend; Yo-Yo Ma, Vilcek Prize winner in music; and Richard Flavell, Vilcek Prize winner in biomedical science, at the Vilcek Foundation Awards Dinner in 2013.

Rick Kinsel (left), executive director of the Vilcek Foundation, with Jeanne-Claude and Christo, winners of the Vilcek Prize in the arts, at the Vilcek Foundation Awards Dinner in 2007.

Hillary Rodham Clinton with me and Marica at a reception at the Metropolitan Museum of Art in 2012. Second from left is Carrie Rebora Barratt, deputy director of the Metropolitan Museum.

Former President Bill Clinton with me and Marica at a private reception in 2013.

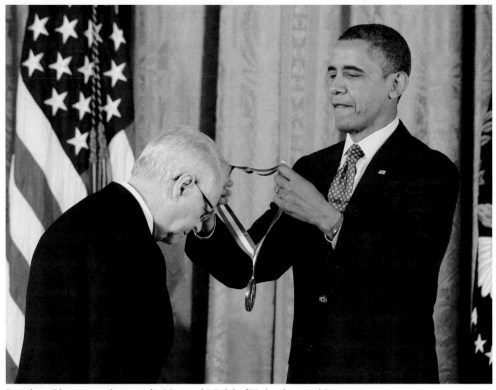

President Obama awarding me the National Medal of Technology and Innovation at the White House on February 1, 2013.

With President Obama following my receipt of the National Medal of Technology and Innovation at the White House on February 1, 2013.

With my former student Douglas Lowy (now acting director of the National Cancer Institute) in 2015 at the celebration of the fiftieth anniversary of my joining NYU School of Medicine.

Marica and I flanked by Kenneth Langøne (left), Chairman of the Board of the NYU Langone Medical Center, and Robert Grossman, Dean and CEO of the NYU Langone Medical Center, in 2015 at the celebration of the fiftieth anniversary of my joining NYU School of Medicine.

some other reasons the publication of Jacobs's preliminary findings of interferon's beneficial effect in MS was greeted with widespread disbelief.

Several years later, the NIH sponsored a much more thorough and more rigorously controlled study that confirmed the worth of interferon treatment for MS. The fibroblast interferon used in the NIH study was produced by a German biotechnology company named Bioferon. Bioferon, in turn, had emerged as the result of a collaboration between a German pharmaceutical company, Rentschler, and the Cambridge, Massachusetts–based biotechnology company Biogen (now Biogen Idec). The DNA construct used for the preparation of the recombinant interferon employed in the MS trial—produced by the Rentschler team—was derived from FS-4 cells supplied by our laboratory. For many years I was advising Rentschler and Bioferon on their interferon program.

Why and how interferon works in MS is still not completely understood. MS is an autoimmune disorder, but the details of the evolution of the disease process are not fully known. The beneficial effect is now thought to be due to interferon's action on the immune system that leads to a suppression of the inflammatory cycle caused by autoimmunity. In any case, the beneficial action of interferon is almost certainly unrelated to its virus-inhibitory activity. Curiously, drugs that block the action of TNF, which, as recounted in the second and third chapters, are highly effective in the treatment of numerous autoimmune diseases, lack effectiveness in MS and may in fact accelerate disease progression.

———

Another center of effort to produce significant quantities of human interferon was a laboratory at the State Serum Institute in Helsinki, Finland, headed by Kari Cantell. (Kari was a participant of the interferon conference I had organized in Czechoslovakia in 1964.) Kari used cultures of white blood cells or "leukocytes"—separated from fresh human blood collected for blood transfusion—to produce "human leukocyte interferon." He initiated these efforts years before we embarked on improving methods of fibroblast interferon production—in the early 1960s—and his laboratory was the first to produce interferon in quantities sufficient for the initiation of limited human trials.

The first patients—mostly terminal cancer patients—were treated with leukocyte interferon in 1969 with no apparent benefit. In the early 1970s Kari and his Swedish clinician colleague Hans Strander were able to prepare enough interferon to treat a larger group of cancer patients, among them patients with a highly malignant and almost invariably deadly bone tumor called osteosarcoma occurring in children and adolescents.

As vividly described in Kari's autobiography *The Story of Interferon: The Ups and Downs in the Life of a Scientist*, the results in the first few patients looked so promising that a decision was made to treat all osteosarcoma patients admitted to the prestigious Karolinska Hospital in Stockholm with interferon. Unfortunately, these clinical studies were not sufficiently rigorous by today's standards.

We now understand that in order to obtain objective proof of a drug's clinical efficacy, trials have to be performed in a "placebo-controlled, double-blind" fashion, which means that only some of the patients are treated with the tested drug and some are given a placebo—essentially a sugar pill or injection of a harmless substance. Neither the patient nor the physician can know whether the treatment administered to any one patient contained the experimental drug or placebo—until the code is broken and the results are evaluated. Some years later it became clear that the favorable results seen in osteosarcoma patients treated with leukocyte interferon were incidental and not attributable to a beneficial action of interferon. However, in the meantime, news of the apparent beneficial effect of interferon in patients with a deadly cancer had spread around the world.

A *New York Times* editorial headlined "New Forms of Therapy" published on July 6, 1973, included the following optimistic prediction of the potential therapeutic worth of interferon:

> Another new therapeutic modality is interferon, the antiviral
> chemical produced in cells. The usefulness of interferon as a
> means of combatting some viral infections has already been con-
> vincingly demonstrated. In recent years the thought has devel-

oped that if interferon is effective against viruses, and if cancer may be a virus-caused disease, then perhaps interferon could be useful against malignant growths.

Others were even less restrained in their enthusiasm. Mary Lasker, who with her husband Albert established the eponymous foundation dedicated to the support of medical research, became an outspoken advocate of allocating public funds to the effort of making more interferon available for use in cancer patients. In 1975 Mary Lasker's friend Mathilde Krim—a Swiss- and Israeli-educated scientist and wife of the influential attorney and businessman Arthur Krim—organized a large meeting at the Memorial Sloan Kettering Cancer Center in New York City with the goal to mobilize public opinion in support of research that would make more interferon available for clinical testing.

As a result of this pressure and publicity, the National Cancer Institute—somewhat reluctantly—allocated one million dollars for interferon research—a large sum in the 1970s—most of it used for the purchase of Kari Cantell's leukocyte interferon. The Lasker Foundation added another million dollars of its own money, and the American Cancer Society allocated two million, the latter representing at the time the largest grant ever awarded by the ACS. The publicity surrounding interferon and its promise as a therapeutic agent for cancer continued for a number of years. On March 31, 1980, *Time* magazine featured the word *INTERFERON* in huge letters on its front page, followed by this pun: *The IF Drug For Cancer.*

Most of my scientist colleagues and I viewed these developments with mixed feelings. On the one hand, we were happy that our area of research was getting attention and that funding was directed to the field. At the same time, many of us worried that the expectations of what interferon therapy would be able to accomplish were being raised to unrealistic levels, and that all this fracas might one day backfire. Indeed, even though interferon did eventually find its way into clinical practice—mainly in patients with chronic active hepatitis B or C, and in patients afflicted with multiple sclerosis—interferon's usefulness in the management of malignancies and in acute virus infections has turned out to be modest.

On balance, it is fair to say that the publicity generated by Mathilde Krim, Mary Lasker, and others helped to advance the field of interferon research. As a result of the publicity, more scientists became interested in the field, and efforts to purify, isolate, and characterize the interferon proteins accelerated, leading to numerous successes within a few years following the 1975 conference. The expectation that interferon would one day become an important therapeutic drug also piqued the interest of pharmaceutical companies and spawned the establishment of numerous biotechnology companies.

Although interferon never became the blockbuster drug prophesized to cure ailments ranging from the common cold to breast cancer, several companies successfully developed interferon products useful for a limited number of therapeutic indications. But there was an even more important outcome. The recombinant DNA technologies developed by these companies for the production of interferon were adapted for use in the production of many other successful biotechnological products, leading to a veritable revolution in therapeutic or prophylactic drug development.

Most of the hundreds of existing pharmaceutical biotechnological products can be divided into two categories. Either, like interferons, they are proteins that are identical or similar to proteins produced by the human body, or they are artificial antibodies called monoclonal antibodies. Here are a few examples of such successful products.

The earliest approved biotechnology product of the first category was recombinant human insulin, introduced in 1982 for use in diabetes patients. Until then most diabetes patients were treated with insulin derived from the pancreas of animals, cows, or pigs, and some people developed allergies to the foreign protein. Introduction of recombinant insulin was followed three years later by the approval of recombinant growth hormone for children with growth hormone deficiency. Also among the early-introduced recombinant DNA products were human erythropoietin, marketed as Epogen or Procrit, used for the treatment of some forms of anemia, and Filgrastim (Neupogen), a recombinant growth factor that helps the body make certain white blood cells, used for the prevention of infections in cancer patients receiving chemotherapy.

Many vaccines were developed only when recombinant DNA technology enabled the production of sufficient quantities of the proteins that form the basis of the vaccine; the best-known examples include the vaccine against hepatitis B, and the human papilloma virus vaccine. Hepatitis B vaccine protects against infection with the hepatitis B virus that can lead not only to acute symptoms of disease but also to ensuing irreversible liver damage and cancer. The more recently developed papilloma virus vaccine protects against uterine cancer and some other cancers caused by papilloma virus infection. (I mentioned the papilloma virus vaccine in an earlier chapter because one of the major contributors to its development is my former student Doug Lowy.)

There are around forty therapeutic monoclonal antibodies approved to date; they are widely used in the management of cancers and other disorders. For example, bevacizumab, better known under its trade name Avastin, is one of the monoclonal antibodies used in some types of metastatic cancers. It works by inhibiting the growth of new blood vessels and thus cutting off blood supply to tumor cells. The same monoclonal antibody, marketed under the name Lucentis, is also used for the treatment of age-related macular degeneration, a common eye disease.

Another example of a successful therapeutic monoclonal antibody is ipilimumab, marketed as Yervoy, used for the treatment of advanced metastatic melanoma and some other cancers. Yervoy is the first-introduced monoclonal antibody of a new class of therapeutic agents called checkpoint inhibitors that work by mobilizing the body's immune system to attack cancer cells. These drugs are called checkpoint inhibitors because they block molecules present on cells of the immune system that act as physiological brakes of the activated immune system. The end result of their action is an enhanced capacity of the body's immune cells to eliminate cancerous cells. Clinical trials at the Memorial Sloan Kettering Cancer Center that demonstrated how Yervoy could dramatically prolong the life expectancy of patients with metastatic melanoma were directed by Jedd Wolchok, a graduate of NYU School of Medicine's MD-PhD program, who received his PhD degree in my laboratory and is now a leader in the important field of cancer immunotherapy. Other monoclonal antibodies that act as checkpoint inhibitors but have targets different from those of Yervoy are now approved for the treatment of metastatic melanoma, and more are being evaluated.

As I recounted in the second and third chapters, the most widely used therapeutic monoclonal antibodies (and therapeutic recombinant DNA products in general) are TNF inhibitors. Remicade was the first TNF inhibitory antibody successfully used to treat humans and the first such therapy approved for clinical use in the US and elsewhere.

Chapter Twelve

Sorting Out Interferon

While it was clear that the procedures used for the laboratory production of human leukocyte and fibroblast interferon were different, initially it was not known whether the resulting products were also different. In other words, was interferon produced in human white blood cells isolated from fresh blood the same as interferon generated in fibroblasts derived from the foreskins of newborn baby boys?

We knew that some of the basic properties—for example the ability of leukocyte and fibroblast interferon to suppress virus replication—were indistinguishable. The easiest way to find out if they were identical or not would have been to analyze the structure of the fibroblast and leukocyte interferon molecules, meaning to determine the composition and sequence of their building blocks—the amino acids. But this wasn't possible at the time because in order to perform such a direct biochemical analysis it is necessary to have the protein molecules in pure form, something that was not accomplished until the very end of the 1970s.

To answer the question of whether human leukocyte and fibroblast interferon were the same, we initially had to rely on indirect methods. Once we started to compare their properties it became apparent that fibroblast and leukocyte interferon did indeed have somewhat different characteristics. For example, under some conditions the biological activity of fibroblast interferon decayed faster than the activity of leukocyte interferon. But that was still not sufficient proof that the two molecules were in fact distinct because the differences could have been due to some other material present in the two preparations affecting interferon's stability. To obtain more direct evidence, my colleague Ed Havell and I joined forces with the laboratory of Kurt Paucker at the Medical College of Pennsylvania in Philadelphia. Like me, Kurt was among the small group of scientists who embraced the study of interferon at an early stage.

We had met at the interferon symposium I had organized in 1964 in Czechoslovakia and Kurt was one of the people I wrote to when we were awaiting our visas in Germany. We became close professional colleagues and personal friends. Sadly, Kurt succumbed to metastatic colon cancer at age fifty-six. When all other treatments had failed, Kurt asked to be treated with interferon, but the interferon injections he received had no beneficial effect.

By the mid-1970s—as a result of progress in the production and partial purification of human interferon—it became possible to generate antibodies to these interferons in laboratory animals. Kurt Paucker's lab produced antibodies by injecting rabbits with human leukocyte interferon and Ed Havell in my lab used rabbits to generate antibodies to human fibroblast interferon. By exchanging sera from these animals we could establish that the antibodies blocked ("neutralized") the activity of the type of interferon that was used for the immunization of animals but failed to neutralize or even bind the other interferon type. Since antibodies are known to recognize differences in protein structure determined by amino acid sequences, our evidence—though still somewhat indirect—was the first convincing proof that the amino acid sequences of leukocyte and fibroblast interferon were different.

In an extension of these studies we joined forces with Sidney Pestka's group at the Roche Institute of Molecular Biology in Nutley, New Jersey. Sid and his colleague Ralph Cavalieri had adapted an ingenious technique that used frogs' oocytes (egg cells) to make human proteins under the direction of messenger RNA isolated from human cells. To explain, messenger RNA is a molecule that "translates" genetic information encoded in the DNA into proteins. Proteins, in turn, are synthesized on specialized cellular structures called ribosomes. Like individual genes, each messenger RNA is unique and usually specific for a single type of protein. In the 1970s it became known that when a messenger RNA extracted from cells is artificially injected into tiny frogs' oocytes, the ribosomes in the oocytes will faithfully translate the injected messenger RNA into the type of protein encoded by that particular messenger RNA.

Armed with this information, we took test tube cultures of human fibroblasts and human leukocytes, exposed them to appropriate treatments to induce interferon production in them, and then extracted messenger RNA from the two cell types. Ralph Cavalieri, in Sid Pestka's lab, injected the extracted RNAs separately into two sets of frogs' oocytes. If the unique specificities of fibroblast and leukocyte interferon were determined by differences at the level of genes and messenger RNAs, then oocytes injected with the two types of messenger RNA would give rise to two distinct interferons. If, however, the differences between fibroblast and leukocyte interferon arose only as a result of the different environments of the two types of cells, then, upon injection into oocytes, the two messenger RNAs would be expected to give rise to identical interferon proteins.

The results of these experiments showed very clearly that the specificities of fibroblast and leukocyte interferon were encoded in their messenger RNAs. Since messenger RNAs are copied from the DNA of individual genes, our results also strongly suggested that the two interferons were encoded by separate genes.

———

Ultimately, direct evidence of the existence of separate genes for leukocyte and fibroblast interferon was provided only a few years later by characterizing the gene sequences encoding the two interferons. A team headed by Charles Weissmann at the University of Zurich, Switzerland, used human white blood cells stimulated to produce interferon according to protocols developed by Kari Cantell in Helsinki. They then extracted messenger RNA from the white blood cells, and made what is called a "complementary DNA," meaning a DNA whose sequence matches that of the messenger RNA. The artificially produced DNA was then incorporated into a larger stretch of DNA and inserted into a strain of bacteria with the hope that the bacteria would recognize the genetic sequence and synthetize human interferon.

Eventually the group in Zurich succeeded in producing small amounts of biologically active human leukocyte interferon in bacteria. This was a major achievement, opening the way to the detailed characterization of the interferon genes and proteins and to the development of efficient methods for industrial-scale production of interferon using recombinant DNA technology. Not only would recombinant DNA technology allow the production of interferon in quantities that were impossible to obtain by the traditional methods, the use of this new technology would also facilitate making interferon proteins available in completely pure form. With time, the new methods of production would open the way to rigorous clinical trials and provide unequivocal answers about the breadth of interferon's therapeutic utility—or the lack of it.

The task of isolating complementary DNA specific for human fibroblast interferon was first accomplished by the Japanese scientist Tadatsugu "Tada" Taniguchi. Tada was a student in Weissmann's lab, but he returned to Japan before Weissmann completed the isolation and characterization of leukocyte interferon complementary DNA. Not wanting to compete with his former mentor, Tada decided to try his luck at "cloning" fibroblast interferon, rather than leukocyte interferon. He extracted messenger RNA from human fibroblasts that were stimulated to produce interferon, using methods we helped to develop.

Tada, working almost single-handedly in Tokyo, managed to complete the first half of the task—the isolation of complementary DNA for human fibroblast interferon, its sequencing, and elucidation of the complete sequence of the human fibroblast interferon protein—before Charles Weissmann had succeeded in completing these steps with leukocyte interferon. Generation of active fibroblast interferon in bacteria took Tada a little longer to accomplish. My laboratory played some role in this feat.

Tada phoned me in the spring of 1980. Although I knew of Tada's work, I had no prior personal contact with him. He was calling from Boston, where he was temporarily working in the lab of Mark Ptashne, a prominent and at the time Harvard-based molecular biologist. There, with help from Ptashne's team, Tada was attempting to find ways to make bacteria "express" (meaning produce) human fibroblast interferon from the complementary DNA he isolated and identified in Tokyo.

Tada told me that they might already have succeeded in their goal, but to verify, would I be willing to help by testing their materials for the presence of interferon activity? In those days, detecting small amounts of interferon in extracts from bacteria was not a trivial matter, and my laboratory was one of the few able to do it. "Sure," I said, "why don't you bring the samples to NYU, and we'll test them."

Tada arrived a few days later, on May 5, 1980, with a large Styrofoam container filled with frozen samples. The reason I remember the date is that it became important in a patent dispute that erupted many years later between Tada's team and a competing group who sought to prove they were the first to show that fully active human recombinant fibroblast interferon could be produced in bacteria.

Tada is not only a brilliant scientist, he is also a worldly and warm human being, passionate about science but also about classical music. We spent hours discussing his project, finding out in the process that we had many common interests outside the realm of science.

With the help of my senior lab technician at the time, Dorothy Henriksen-DeStefano, we set up assays to determine if the samples brought from Boston contained interferon activity. Tada returned to Boston before the tests were completed. When they were ready to be analyzed, the results showed that the bacterial extracts Tada brought to New York did indeed contain detectable amounts of activity characteristic of human fibroblast interferon. I immediately called Tada to congratulate him on his important accomplishment. Thus began another long-lasting professional relationship and personal friendship.

By the early 1980s, the pioneering studies of Charles Weissmann and Tada Taniguchi—together with contributions from many other laboratories and individuals—revealed a great deal of information about interferon genes and proteins. These studies showed that, as predicted, leukocyte and fibroblast interferon are indeed encoded by separate genes. But there were unexpected surprises. Whereas in humans there is a single gene for fibroblast interferon, it turned out that leukocyte interferon comprises a family of twelve related genes and proteins. Although they are all very similar to one another in their structures, subtle functional differences exist among the different interferon family members.

It also became clear that leukocyte and fibroblast interferon, though distinct, share some structural homology, which explained why they were similar in their biological activities. By comparing their sequences, it was possible to conclude that the fibroblast and leukocyte interferon genes arose during evolution from a common ancestral gene roughly 250 million years ago. These differences have some practical implications because the therapeutic uses of fibroblast and leukocyte interferons are different.

Fibroblast and leukocyte interferons are now referred to as type I interferons—to distinguish them from some other forms that are called type II and type III interferons. In the 1980s, the terms "leukocyte" and "fibroblast" interferon were replaced with new, less onomatopoetic designations "interferon-alpha" and "interferon-beta," or IFN-alpha and IFN-beta for short.

Hence, some twenty-five years after the first description of interferon by Isaacs and Lindenmann, interferon proteins and genes had finally been defined in strict scientific terms—quashing any possible lingering doubts as to whether these molecules truly exist. Soon it would become possible to produce virtually unlimited quantities of interferon using recombinant DNA techniques, thus for the first time providing enough material for extensive scientific studies and clinical trials. From these clinical trials we have learned that IFN-alpha can be used to treat chronic infections with hepatitis B and C viruses, and that therapy with IFN-beta can slow the progression of multiple sclerosis. IFN-alpha has also been approved as adjuvant therapy for some forms of malignancies, although the actual use of interferon in cancer patients is now relatively rare. Equally important, as a result of the molecular characterization of IFN-alpha and IFN-beta, we now have a much better understanding of the intricate roles they play inside our bodies.

———

As my professional work was becoming better known, colleagues at other institutions started asking me if I would consider job offers. I had no reason to leave NYU, but found it reassuring that others valued my work. One day in the late 1970s I received a phone call from a Swiss colleague at the medical school of the University of Geneva. Would I be interested in visiting Geneva to explore the possibility of becoming the chairman of their Department of Microbiology? Because I was far from fluent in French—the official language at the University of Geneva—I thought it was very unlikely they would offer me the job. Nevertheless, wouldn't it be interesting to visit Geneva and explore how a reputable medical school in Europe functions? I decided to accept the invitation.

Only a few days after my return to New York I received a phone call from Geneva: they decided to offer me the position and an official letter spelling out the details of the offer was on its way. I was quite stunned.

To me the prospect of moving to Geneva was not without appeal. A civilized, cosmopolitan city, close to France and Italy. The professional environment and opportunities seemed excellent. Another consideration was that New York in the 1970s had serious problems—crime was rampant, scores of people were moving out of the city, and there was a fiscal crisis. Would it perhaps make sense to leave before things got even worse? But Marica was passionately opposed to moving. "One emigration in a lifetime is enough," she said. "Now that I've been able to adjust to life here and have achieved some degree of professional success I am not going to throw it all away and start from scratch." I sent word to Geneva that I was unable to accept the offer.

I received other job offers within the US, and each time Marica's opinion played an important role. "But how am I going to find a position comparable to my Met Museum job?" she would ask. We stayed in New York City.

Years later, in 1989, an opportunity arose for me to be considered for the position of chairman of my own department at NYU School of Medicine. Dr. Salton, who had hired me nearly a quarter of a century earlier, was ready to step down as chair. Saul Farber, the dean of our medical school, decided to fill the position with an internal candidate. He appointed a search committee and asked its members to recommend an appointee from among three preselected candidates. I was one of them.

Another candidate was Claudio Basilico, an Italian-born virologist who came to the California Institute of Technology to study with Renato Dulbecco, winner of the Nobel Prize for pioneering work on tumor viruses. At the time, Claudio was a professor in NYU's Department of Pathology. There was a third candidate, a prominent scientist with strong research credentials in microbiology, but the general feeling was that the final choice would be between Claudio and me.

A description of Claudio's personality could easily fill a separate chapter. Over six feet tall and sturdily built, his physical presence alone commands attention and respect. To say that he has strong opinions is an understatement; if there is one dissenting voice during a group discussion, it is likely to be Claudio's. Yet his personal magnetism easily smooths out his rough edges.

Before the search committee held its formal meeting, I had been approached by its chairman. Would I be ready to accept the post of chairman of microbiology if I were selected? Claudio was not only a close colleague but also a personal friend. I knew that he was eager to assume the chairmanship. Whereas I was not so sure if I wanted to take on more administrative responsibilities. My main goal was to be able to devote as much time as possible to my research, and becoming chairman would inevitably reduce the time available to do science.

There were other reasons that made me hesitate. I have little patience for listening to peoples' complaints—a significant component of a chairman's chores. All in all, I came to the conclusion that Claudio would be more effective as chair. I withdrew myself from consideration and endorsed Claudio's candidacy. Claudio seemed grateful for my decision. In due course Claudio was selected and his choice approved by Dean Farber.

After the decision had been announced, I ran into the outgoing chairman, Milton Salton. "I understand you were not interested in taking over my job," he said. From the tone of his voice I could not tell if he was relieved or disappointed.

In 1979 Ed Havell left my lab to assume a more permanent position at another institution and a Hong Kong–born, Columbia University–trained protein chemist, Yum Keung "Y.K." Yip, joined our group as a postdoctoral fellow. Y.K.'s initial project was to purify and characterize human fibroblast interferon. After working on the project for a few months we learned that another group had succeeded in achieving the same goal. Reaching the finish line in second place may be an honorable achievement in sports, but not in science. This is a cruel fact many scientists encounter repeatedly during their careers. The experience can be especially traumatic for younger scientists who have yet to establish their reputations, but getting scooped after investing a great deal of effort into a project is always a blow for seasoned investigators as well.

We put our heads together to figure out what to do next. Despite Y.K.'s initial reluctance, we agreed to attempt the purification and characterization of another variety of interferon, called type II or immune interferon, which is now usually referred to as "interferon-gamma" or IFN-gamma. This protein was first described in the 1960s as an interferon-like substance produced by a type of white blood cells called lymphocytes. IFN-gamma suppressed the multiplication of viruses in a manner similar to leukocyte and fibroblast interferon. But in many other respects IFN-gamma was different and some people questioned if it was appropriate to refer to it as interferon. Only purification of the protein and cloning of its gene could resolve the question of how, if at all, IFN-gamma was related to the type I interferons.

The procedure employed to produce IFN-gamma was different from the techniques used to make fibroblast or leukocyte interferon. Similar to the process used for the production of leukocyte interferon, preparation of IFN-gamma required white blood cells derived from fresh human blood, but the way the cells were then manipulated to produce IFN-gamma was quite different from Kari Cantell's method for the production of leukocyte interferon. Specifically, for the production of leukocyte interferon, the cells were exposed to a virus, whereas to induce IFN-gamma white blood cells were treated with a mixture of two chemicals.

Once we solved some technical problems, such as finding a suitable supply of white blood cells, Y.K.'s effort to purify IFN-gamma progressed successfully. We managed to obtain small amounts of pure IFN-gamma protein and to show that—contrary to earlier published studies—its molecular size was similar to that of the type I leukocyte and fibroblast interferons. We also showed that there were two forms of IFN-gamma that differed in the amount of carbohydrate linked to the protein.

We described our results in two papers published in prestigious journals and we felt good about it. But ultimately this success turned out to be a case of winning a battle but losing the war. An achievement more significant than purification of the IFN-gamma protein was the cloning of its complementary DNA, determination of the DNA sequence encoding it, and artificial synthesis of the IFN-gamma protein by methods of recombinant DNA. In collaboration with my Japanese colleague and friend Tada Taniguchi, we made an effort at achieving these goals, but we were beaten by a group of colleagues at the legendary biotechnology company Genentech in South San Francisco, led by one of the most highly skilled molecular biologists in the world, David Goeddel. The blow to our egos was only slightly tempered by Dave's public acknowledgment that elucidation of some fundamental characteristics of IFN-gamma by Y.K. and me facilitated their task.

Parenthetically, though cloning complementary DNA for IFN-gamma, determining its complete sequence, and producing the IFN-gamma protein by recombinant DNA techniques represented a major scientific achievement, IFN-gamma has not found significant therapeutic applications in medicine. It has been tried in several diseases, but failed to show a convincing benefit, partly because of its toxic side effects. Hence the admirable effort invested by Genentech into the study of IFN-gamma has not turned into a commercial boon. No need to feel sorry for Genentech, though; they have made numerous other discoveries and have introduced many successful products. Genentech is now completely owned by the Swiss pharmaceutical giant Roche. In the final phase of Genentech's acquisition, in 2009, Roche paid $46.8 billion for the 44 percent share of Genentech that it did not already own.

Even though we lost the race for the cloning of IFN-gamma, the work we had invested in its study brought us one unanticipated benefit. It was through our original interest in IFN-gamma that my attention was drawn to another immune modulator termed "tumor necrosis factor," now usually identified only by its acronym TNF. TNF was to become the focus of my professional work for some twenty-five years. This effort also led to an accomplishment for which my laboratory would become best known to the outside world—our contribution to the development of the therapeutic drug Remicade, the story with which this book begins.

PART SIX | In Search of a Legacy

Chapter Thirteen

Philanthropy

The FDA approved Remicade for use in Crohn's disease in August 1998, and for patients with rheumatoid arthritis in November of the following year. The very first quarterly royalty payment I received from NYU in the spring of 1999 was only slightly less than my annual professor's salary. For the first time in my life I earned money for which we had no immediate need. Money, in fact, that already exceeded the small daydreams we had of taking taxis and eating out more often, or buying a few more clothes.

The royalty payments continued and showed a gradually rising tendency throughout 1999 and 2000. But would Remicade continue to be successful? At the time it was still far from clear. Could the drug sustain the health improvements shown in the short-term use for Crohn's disease and rheumatoid arthritis? What would be the cumulative side effects if Remicade were to be used for months or years?

It was not unprecedented that a drug would be approved and the approval subsequently revoked because of issues that did not become apparent during the clinical trials. Centocor's Centoxin was approved in Europe for the treatment of sepsis, but when the FDA had ruled that it did not produce a significant benefit, the European approval was withdrawn. A better-known example is Vioxx, an anti-inflammatory drug brought on the market with great fanfare in 2001 by the pharmaceutical giant Merck & Co., only to be "voluntarily withdrawn" by the manufacturer in 2004 because of an increased risk of heart attacks among the drug's users. It was very possible that Remicade might meet a similar fate.

Toward the end of the year 2000 I was approached by representatives of Drug Royalty, Inc., a company based in Canada that earns its keep by purchasing rights to future royalty payments for pharmaceutical products in exchange for an up-front cash lump sum. Would I be interested in selling them a portion of my future royalty income I was entitled to by the 1984 licensing agreement between NYU and Centocor? Until Drug Royalty approached me I had no idea that future royalties could be bought and sold.

I thought about it. If Remicade did well in the future, I would lose money, but if Remicade did not do well in the marketplace, or worse, had to be withdrawn from the market because of side effects, I would at least retain the cash received from Drug Royalty. I viewed the transaction as a kind of insurance policy and decided to sell a small fraction of my future royalty entitlement. It was a complicated transaction. I needed approval from NYU and I needed business and legal advice, the latter provided by Peter Ludwig, the patent attorney and friend who had also helped to negotiate the original NYU-Centocor license agreement.

In retrospect, the deal has turned out to be the worst business transaction of my life. Drug Royalty has earned its money back many times over. But I don't mind. In fact, I much prefer to have lost money than to have needed the "insurance" because the latter would have meant that Remicade did not become successful. How things have turned out was a better alternative for everyone involved.

———

At about the same time as I was negotiating with Drug Royalty about the sale of a portion of my future royalty stream, Marica and I were trying to solve the enviable problem of what to do with the additional moneys flowing each quarter into our bank account when I was paid royalties received by NYU from Johnson & Johnson for sales of Remicade. We used a small portion of the additional earnings to make our lives more comfortable, and to be more generous to our friends and family. But if the inflow of funds were to continue and perhaps even grow, we realized that, for the first time, we really needed to come up with some financial planning.

We were not interested in drastically changing our lifestyle. We did not need to own a second home. We liked the freedom of spending our weekends wherever we liked (usually in New York City enjoying the city's many cultural offerings), to travel to different destinations, and not to feel pressured to go week after week to the same country house. Neither did we yearn for a private plane or a yacht. In fact, like numerous other New York City dwellers, we had decided many years ago to give up ownership of our car, and we continue living a carless life. Using public transportation or taxis in the city and renting cars for out of town excursions suits us fine.

Gradually the thought occurred to us that we could use the unexpected windfall for something that would benefit society. Start a foundation perhaps?

In our first planning session with Paul Roy, an attorney expert in the area, we spent more than an hour learning the basics about the logistics of starting private charitable foundations. When we told Paul that the funds we were considering using to establish the foundation would be derived from royalty payments, he told us that instead of funding the foundation with cash, we could donate a specific segment (legally termed "an undivided fractional interest") of my future royalty stream. Paul warned that such a donation would be irreversible; once I made the commitment to assign to the foundation a certain percentage of the rights to my future royalty stream, that decision could never be changed again.

The option to donate to the foundation a fixed portion of my future royalty income, rather than making one or more donations of cash, appealed to us. We had no good idea what the royalty payments would amount to over the years. Some Wall Street analysts had published forecasts of anticipated Remicade sales, but their reliability was anybody's guess. Gifting to the foundation a set percentage of the royalty income I was entitled to receive from NYU throughout the lifespan of the patents would make the guessing game unnecessary. If future Remicade sales did well, the foundation would benefit; if they did not do so well, the foundation would end up with less. We decided to donate a segment of the royalties.

One of the next steps was to set up the foundation and to choose its name. We were too self-conscious to use our own name, so instead the foundation was named after my mother: the Friderika Fischer Foundation, Inc., formed on December 1, 2000.

And what would be its major mission? The royalties were generated by the sales of a drug used for the treatment Crohn's disease and rheumatoid arthritis, two chronic inflammatory autoimmune disorders, and more research was needed to develop a better understanding and additional treatments for these and related diseases. It seemed logical to direct the main activities of the newly established foundation toward the support of research in the same field. And so the bylaws stated that one of the main purposes of the foundation was to "engage in research pertaining to the development of treatments for chronic inflammatory autoimmune diseases."

Now the foundation had a name, legal status as a corporation, and bylaws with a stated mission. Soon thereafter, the IRS officially approved its status as a tax-exempt charitable organization, a 501(c)(3).

——

I served as the president and treasurer and Marica as vice president and secretary. The official address of the foundation was our private residence, and a small built-in desk in the kitchen area alternated as foundation office space. According to the bylaws, Marica and I needed to appoint a board of directors, the governing body of the foundation. We invited three people. Bruce Cronstein, a physician and scientist colleague from NYU with expertise in rheumatoid arthritis and other chronic inflammatory diseases, and two of Marica's former colleagues from the Metropolitan Museum, Jennifer Olshin and Rick Kinsel.

As planned, starting around 2001 and continuing for several years, the foundation dispensed money in support of two science projects, both conducted at NYU School of Medicine, that were broadly focusing on research in inflammation and possible avenues to treat inflammatory disorders. We also provided small grants to other science-related programs, including fellowships for young people attending meetings of the International Cytokine Society and funds in support of a scientific conference devoted to the TSG-6 protein, the molecule that was discovered in my lab.

Marica, Rick, and Jennifer argued that, in addition to providing funds to science-related projects, our foundation should consider supporting projects in the arts. Our first grant in the arts field, in 2002, went in support of the apprentice program at the Santa Fe Opera—a logical choice because Marica and I spent many summer vacations in Santa Fe, over the years we enjoyed seeing many interesting opera performances there, and the SFO apprentice program had had a long tradition of nurturing future top opera singers.

The Friderika Fischer Foundation was not our only philanthropic endeavor and, in financial terms, not even our largest one. As my royalty checks continued to increase, I made the decision to make a major gift to NYU School of Medicine. I felt strongly, and Marica concurred, that NYU deserved to be the major benefactor because NYU's School of Medicine had the courage to offer me a faculty position when I arrived here as a penniless refugee. Over the years, NYU provided me with the freedom to develop my research program, offering the supportive environment I needed, never interfering with my choices of research topics or collaborators. In the pre-Remicade days I would often comment that if it weren't for the fact that I needed my salary to survive, I would have gladly paid NYU two or three times the amount they were paying me—for the privilege of being able to do my job. Little did I know that one day I would be able to do that—and more.

By 2002 it was clear that Remicade was going to be a success. I consulted with Paul Roy, the attorney who helped us set up our foundation. What would be the best mechanism to make a major gift to NYU? Paul recommended that I again make an irrevocable gift of a portion of my future royalty income to NYU. I agreed.

A "Deed of Gift and Agreement" was signed by me and Dr. Robert Glickman, the dean of NYU School of Medicine at the time, a few days before the end of 2002. The donated royalty income was to be apportioned to four different programs: to establish an endowed chair in the Department of Microbiology, support research in the field of microbial pathogenesis, create endowed fellowships for graduate students and post-doctoral fellows in the Department of Microbiology, and to support research and recruitment to the Department of Otolaryngology.

Most of the support was designated to benefit the Department of Microbiology, my professional home since 1965. I was pleased that my longtime friend and colleague Claudio Basilico became the first Jan T. Vilcek Professor of Molecular Pathogenesis. Upon taking over the chairmanship of Microbiology in 1990, Claudio revitalized the department, hiring a score of younger faculty members. Though intimidating at times, especially to students not accustomed to his authoritative manners, Claudio became an effective and respected leader. He stepped down as chair in 2014.

I designated NYU's Department of Otolaryngology as an additional beneficiary of the donated funds because Marica had been their longtime patient. A few years after our moving to New York, Marica developed Ménière's disease, an inner-ear disorder that affects hearing and balance. During acute attacks of the disease Marica suffered from debilitating nausea and vertigo. When medications and several surgical interventions failed to help, she had to undergo "labyrinthectomy," a surgery that destroys one of the two balance centers. An inevitable side effect of labyrinthectomy was the loss of hearing on the operated side. The surgery, though drastic, restored Marica's ability to live a normal life and we are grateful to the Department of Otolaryngology and its former chairman Noel Cohen, who performed the surgery, for the care she had received.

When I donated a portion of my future royalty income to NYU in 2002, the ultimate value of the gift was, of course, not known. That the value of my donation exceeded the original expectations started to become clear about three years after the effective date of my donation. Dean Glickman called me to his office to tell me that we needed to amend the original 2002 agreement because the programs designated to benefit from the gift would soon become fully funded and there were no instructions specifying what to do with the additional royalties that were expected to accumulate for about twelve more years.

The signed amendment to the original agreement stipulated that once the full funding levels of the four programs were reached, NYU Medical School would use any additional income from my gift to "strengthen the basic sciences programs at the University's School of Medicine." The agreement went on to list specific examples of such uses.

This language reflected my strong belief—partly based on my own experience—that advances in medicine, including progress in the treatment of diseases, are most likely to emerge from advancements in basic science, rather than from so-called applied research. In the last decade it has become fashionable to promote "translational medicine"—the use of new scientific developments for the furtherance of advances in medical practice, with the idea that this trend will accelerate the transfer of knowledge "from bench to bedside." In principle, I support these efforts, but it is important to keep in mind that without continued generous support of research in the basic sciences, there soon would be nothing left to "translate."

The history of science has taught us that the path from basic discoveries to clinical applications is rarely linear. Already in 1921 Marie Curie had the following to say about her and her husband's historic discovery: "We must not forget that when radium was discovered no one knew that it would prove useful in hospitals. The work was one of pure science. And this is a proof that scientific work must not be considered from the point of view of direct usefulness."

Not so long ago, research into embryo development of fruit flies (*Drosophila melanogaster*) led to a strikingly new interpretation of the mechanism of immune responses to microbial invaders. Christiane Nüsslein-Volhard and her colleagues in Tübingen, Germany, discovered a gene they termed "Toll." In drosophila this gene is important for the formation of the so-called dorsal-ventral axis during embryogenesis. Later it was discovered that receptors similar to Toll, the "Toll-like receptors," play a central role in immunity against infections in higher organisms, including *Homo sapiens*. In the address delivered on her receiving the Nobel Prize in Physiology or Medicine, Nüsslein-Volhard said she had had no inkling that her research in drosophila would turn out to be important in medicine.

When, in the early 1980s, we witnessed the emergence of the AIDS epidemic, nothing at all was known about the cause of the illness. The name given to the condition, "acquired immunodeficiency syndrome," reveals that at the time it was not even known that we were dealing with a transmissible infectious disease. People of my generation will remember that early on one of the leading theories was that AIDS is caused by a drug used by young gay men as a sexual stimulant. Credit for the relatively rapid elucidation of the viral etiology of AIDS—and for the fact that within a few years scientists were able to develop effective drugs that can control HIV infection—goes to information derived from earlier studies of a family of viruses called "retroviruses." The goal of these earlier studies was to explain how retroviruses cause tumors in animals, for example chickens, but at the time it was not clear at all if these findings would have any relevance for human disease. HIV, the causative agent of AIDS, turned out to be a member of the retrovirus family. Without extensive information on retroviruses, acquired in the course of the earlier basic research, it would likely have taken many more years to elucidate the cause of AIDS and to develop effective treatments.

About a decade ago, in an effort to speed up the application of advances in medical research toward the development of new cures, the federal government initiated a number of programs that provide incentives to scientists to engage in translational research. NIH established a large number of Clinical and Translational Science Centers around the country and committed hundreds of millions of dollars to their support. Since the overall budget of the NIH and other agencies supporting medical research in the country has been stagnant in recent years—in fact going down in real dollars—the funds used for this initiative were taken from the pool of money originally earmarked for basic science research.

As a result, it is now often easier to obtain funding for a translational research project than for first-class basic research. New professional journals with the word "translational" on their mastheads are proliferating, along with new professional societies devoted to translational medicine. In fact, I was recently honored by one of these new societies, the Association for Clinical and Translational Science.

The development of Remicade was a prime example of translational science in action, accomplished at a time when the word "translational" had not yet become fashionable. Yet I believe that the balance has tipped far in the translational direction. As microbiologists Ferric C. Fang and Arturo Casadevall aptly put it in an editorial a few years ago:

> It will be critical not to allow our impatience for translational applications to skew resources and researchers away from the open-ended exploration of the natural world that has provided the foundation for so many translational successes and remains as essential as ever.

By 2005, I was ready to make additional donations to NYU. By then Marica and I felt that giving another major chunk of my future income would not jeopardize our already very comfortable lifestyle. We went on to set up two trusts, of which NYU School of Medicine was the sole beneficiary. The trusts were funded with additional portions of my future royalty stream.

In August 2005, a front-page story in the *New York Times* about my gift bore the headline "Research Scientist Gives $105 Million to N.Y.U." Dr. Glickman was quoted as saying: "It's the largest gift in the history of the medical school, and I think the largest by any faculty member to any school anywhere."

At the time I thought the $105 million estimate of the value of my gifts to NYU Medical Center may have been a trifle inflated, but in fact it once again turned out to be overly conservative. Royalty payments to NYU are still accruing, and are likely to continue accruing until the scheduled expiration of the last patents in 2018, so the final value of my gift is still not known. However, through the end of 2014 the combined value of my 2002 and 2005 gifts to NYU has already exceeded $120 million. The additional funds have been used to establish two more endowed professorships, one in microbiology and one in otolaryngology, the latter in Marica's name. The rest of the funds from the 2005 gifts were designated for the renovation of laboratory spaces, hiring of new basic science faculty, and a score of other projects.

The endowed professorship I established in microbiology in 2005 was named after Albert B. Sabin, a graduate of NYU School of Medicine, an early pioneer of virus research and—together with Jonas Salk—a key scientist credited with the development of a vaccine against poliomyelitis. I had a deeply personal reason for honoring Albert Sabin's memory because he played an important role in setting me on the path of my interferon research when I had met him in Bratislava in the late 1950s. Robert J. Schneider, my colleague in the Microbiology Department, has been named the first Albert B. Sabin Professor of Microbiology and Molecular Pathogenesis.

Susan B. Waltzman, professor and codirector of the Cochlear Implant Center at NYU, has become the first Marica F. Vilcek Professor of Otolaryngology. Her work has helped to restore hearing to hundreds of children and adults.

Since 2005, we have made several additional gifts to the medical center, now called NYU Langone Medical Center, in recognition of a landmark $200 million gift by Kenneth Langone, the chairman of the medical center's board. (Ken Langone has kindly said that it was my gift that inspired his generosity.) One donation, made after consultation with our current dean and CEO, Dr. Robert Grossman, that has given Marica and me tremendous pleasure, is the endowment of merit scholarships for talented medical students who, without the scholarship, would either not be able to attend medical school at all or could not attend NYU School of Medicine. Ten students—all of whom we have befriended—have benefitted so far.

Another significant donation helped to refurbish a residence hall for NYU medical students. Upon making this pledge, Dean Grossman asked whether Marica and I would agree to have the residence building named after us. We thought about it and without much hesitation said yes.

Another organization that has benefitted in a significant way from our newly found philanthropy is the Metropolitan Museum of Art, Marica's former professional home for thirty-two years. Our most significant support has been earmarked for the establishment of two endowed curatorships in MMA's American Wing. Two outstanding women art historians have been chosen as Marica F. Vilcek Curators in American Art: Amelia Peck, whose expertise includes American textiles, furniture, and interiors, and Thayer Tolles, who specializes in American sculpture.

Our philanthropic gifts landed Marica and me in some unexpected company. *The Chronicle of Philanthropy* regularly publishes an annual list of the topmost American philanthropists of the year. The list for 2005—the year when I made the pledge to NYU, at the time estimated to be worth $105 million—was published jointly with the online *Slate* magazine. Marica and I were featured among the top fifteen most generous American philanthropists of the year. Most other donors on that list were billionaires whose names were—and are—readily recognizable by the public.

It is worth pointing out that in 2005 when we made the pledge our *real* net worth (not counting future royalty income) amounted to only a fraction of the $105 million. In effect we gave away money before it could even properly be ours. To date, we have given away an amount that is much greater than all of our personal assets combined. Marica and I shake our heads in disbelief at where our life's journey has brought us.

Chapter Fourteen

Vilcek Foundation

About two years into the life of the Friderika Fischer Foundation we started to have second thoughts about the wisdom of naming the foundation after my mother. Very few people knew who Friderika Fischer was or—when told—appreciated the symbolic significance of the foundation's name even though my mother, after all, was my earliest example and supporter in becoming a physician-scientist. Also, perhaps, by this time Marica and I had become less self-conscious about seeing our name on the masthead of the foundation and had come to understand that the foundation will do better if there are human faces behind it. And so, after some deliberations and discussions with our board members, in January 2004 the foundation was rebaptized with its present name, the Vilcek Foundation.

Plans for a more important change in the mission of the foundation started to percolate in our heads at about the same time. Although we were generally satisfied with the progress of the two research projects the foundation was supporting at NYU, we felt we were not breaking much new ground. The royalty payments were growing somewhat from year to year, but we were still a relatively small foundation, with an annual budget of well under one million dollars. Knowing that there were foundations supporting biomedical research with assets hundreds or thousands of times greater than ours, we wondered if it wouldn't make sense to find a more unique niche. Both Marica and I wanted the foundation to do something that would make a difference.

We put our heads together. Eventually, our discussions boiled down to this: Both of us are immigrants. I am a biomedical scientist and Marica is an art historian. Could we perhaps build the foundation program around our combined experiences? Immigrants make enormous contributions to the sciences and the arts in the United States, but are these contributions fully known and appreciated by the general public? Also, as a result of the tragic events of September 11, 2001, the generally friendly attitude toward immigrants Marica and I had enjoyed when we had come to the US had started to sour. By raising public awareness of the contributions of foreign-born scientists and artists, we could try to counter the negative sentiments toward immigrants that seemed to be brewing.

The idea of establishing a prize program for foreign-born scientists and artists was first aired at our annual board meeting in 2003. It would take another three years before the prize program became a reality.

Soon we realized that the amount of administrative work needed to run the foundation was becoming too much for me and Marica to handle. In the fall of 2003, Marica and I persuaded Rick Kinsel—Vilcek Foundation board member, and Marica's former colleague from the Metropolitan Museum—to assume the full-time position of executive director of the Vilcek Foundation.

In 2005 the board approved the launch of the Vilcek Foundation prize program. The first prizes were slated for award in the spring of the following year. After some deliberations, we agreed that there would be two prizes given annually, one in biomedical science, one in the arts. The prizes would be awarded to persons born outside the US who made major contributions to their fields while living in the US. We came up with a list of desired characteristics of the prizewinners: they should be recognized leaders in their fields, individuals known for their personal integrity, and persons who are still active and productive, likely to continue making important contributions.

Because of the large pool of outstanding foreign-born biomedical scientists active in the US, we decided that the prize in biomedical science would be given every year—a decision that undoubtedly also reflected my own personal bias. In contrast, we decided that the particular arts discipline would change from year to year. In fact, between

2006 and 2015 Vilcek Prizes have been awarded in ten different fields of the arts: fine art, architecture, classical music, filmmaking, culinary arts, literature, dance, popular music, design, and fashion. Separate juries were appointed to select the prizewinner in each of these fields. Slated for 2016 is theater.

Winners were awarded $50,000 each in the first years, and starting in 2011 the awards were raised to $100,000. In addition, we decided to give each winner a unique, personalized trophy that was a work of art in its own right. The foundation engaged Stefan Sagmeister, himself an Austrian immigrant to the US and designer of international standing, to come up with the concept of the trophy. Stefan's design is a twelve-inch spire produced by a rapid prototyping (3-D printing) process in which thin layers of photopolymer resin are laser-impressed one on top of the other. The final product is unique and elegant, with each trophy featuring the recipient's name in raised lettering.

The caliber of the first prizewinners has set a high standard for all subsequent years.

The first scientist we selected was Joan Massagué, a prominent Spanish-born cancer researcher at the Memorial Sloan Kettering Cancer Center in New York City. He was chosen mainly for his work on the control of cell growth and cell fate by an important family of growth factors called TGF-beta. The choice of Joan has turned out to be fortuitous. Not only has Joan continued to do spectacularly well in his scientific career (he has since become director of the Sloan Kettering Institute), he has also become an enthusiastic supporter of our foundation, a member of the jury for the selection of our biomedical science prizewinners, and a member of our board of directors.

The arts field we picked for recognition in the first award year was fine art (painting and sculpture), with the husband-and-wife team of Bulgarian-born Christo and Moroccan-born (of French parents) Jeanne-Claude selected as winners. It was indeed appropriate and timely to honor Christo and Jeanne-Claude only one year after their spectacular installation of *The Gates* in New York's Central Park, a project consisting of "7,503 vinyl gates, with free-flowing nylon fabric panels, anchored to 15,006 steel bases on twenty-three miles of walkways." (Thank you, former mayor Michael Bloomberg, for making this stunning project possible!)

***

Our first award ceremony was on March 21, 2006, at the ballroom of the Mandarin Oriental Hotel, overlooking Columbus Circle. As we had no support staff, Marica, Rick, and I, with the assistance of a few personal friends, did virtually everything. The night before the award dinner we worked late into the night assembling nametags and preparing place cards for over two hundred invited guests. We were mightily nervous.

But in the end all—or almost all—went well. Agnes Gund, the philanthropist, art patron, and former president of the Museum of Modern Art, kindly agreed to introduce Christo and Jeanne-Claude. In her speech, Ms. Gund praised the accomplishments of the artist couple, extolling the virtues of *The Gates*. At one point in her speech she said that the installation of *The Gates* could be completed so efficiently because Christo and Jeanne-Claude had several hundred devoted volunteers who helped with the installation. At that point Jeanne-Claude, who was sitting at a front table near the podium, interrupted loudly, "No, we don't use volunteers, we pay everyone for their work."

In acknowledging their award, Christo and Jeanne-Claude recalled their early days in New York City when they both were illegal immigrants (with flame-haired Jeanne-Claude doing most of the talking), and they were scared even of traffic cops, fearing that they would arrest them, when in fact the cops couldn't have cared less about their status. They spoke, too, of their love of New York City, and, in particular, how they—two people from different countries, with different mother tongues—can feel perfectly at home here. Sadly, Jeanne-Claude died three years later at age seventy-four from complications of a brain aneurysm. Christo energetically continues to lobby for their environmental art installations in different parts of the world.

The science prizewinner, Joan Massagué, was introduced by Harold Varmus, cowinner of the Nobel Prize for Physiology or Medicine, and, at the time, president of the Memorial Sloan Kettering Cancer Center. Harold showed slides with photos of graduate students and postdocs from his lab, who hailed from many different parts of the world. (Coincidentally, Harold's colleagues were posed in Central Park among *The Gates*.) He stressed how important the contribution of foreign-born scientists was not only at the top level, but also among the younger professionals who do much of the bench work in the laboratories.

Joan recalled in emotional language how he came to the US from Barcelona for postdoctoral training, fully intending to return to his native Catalonia after the completion of the fellowship. But he had changed his mind when he was offered professional opportunities that could not be matched in Spain. He pointed out that there is no country in the world so conducive to doing world-class science as the US, especially for the foreign-born, because their contributions are welcomed and recognized.

In parallel with the development of the Vilcek Foundation prize program, we have been collecting documentation on the important contributions made by foreign-born professionals to this country. Given my own background, it is not surprising that much of the documentation relates to the role of foreign-born scientists, especially biomedical scientists. Here are some highlights of our research.

Between 1901 and 2014, ninety-eight scientists working in the United States were awarded the Nobel Prize in Physiology or Medicine. Of the ninety-eight winners, thirty-three, more than one in three, were born outside the United States. In the last eight years alone, seven out of the twelve winners of the Nobel Prize in Physiology or Medicine recognized for work done in the United States were not born in this country.

Other measures of the contribution of foreign-born biomedical scientists tell the same story. Among the most coveted grants a US mid-career biomedical scientist can receive are the Howard Hughes Medical Institute Investigator awards. They are as prestigious as the MacArthur

Foundation "genius awards," and their financial value is much greater. In 2015, twelve of the twenty-six HHMI Investigator award recipients (46 percent) were foreign-born. The actual percentage of foreign-born awardees may be even higher because we were unable to establish with certainty the birthplaces of six awardees.

Another example: Research carried out by Stuart Anderson at the National Foundation for American Policy shows that a very high percentage of cancer researchers at the most prominent cancer research centers in the United States are foreign-born: 62 percent at the University of Texas MD Anderson Cancer Center in Houston and 56 percent at the Memorial Sloan Kettering Cancer Center in New York City.

The enormous contribution of foreign-born scientists is by no means limited to biomedical science. Here is one example: The Simons Foundation, a New York City–based private foundation, grants generous long-term "Investigator Awards" to highly accomplished scientists in the fields of mathematics, theoretical physics, mathematical modeling of living systems, and theoretical computer science. In 2015, nine of the sixteen US–based Simons Foundation Investigator Award recipients (56 percent) were born outside this country. In 2014, eleven of fifteen Simons Foundation Investigators (73 percent) were foreign-born.

Perhaps most indicative of the contribution immigrants are making to the *future* of science in the United States are the findings reported by Stuart Anderson that analyzed the birthplace of parents of high school student finalists of the 2011 Intel Science Talent Search. Seventy percent of the forty finalists of this enormously competitive and prestigious competition had parents born outside the United States!

According to census data, the foreign-born population currently represents around 13 percent of the total population of the United States—almost one-third of that being unauthorized "illegal immigrants." It is clear that the contribution of legal naturalized citizens and green card holders to the scientific enterprise in our country greatly exceeds their numerical representation in the population.

A final thought. Immigrants make important contributions at multiple strata of society, from low-skilled workers taking on jobs that the native-born won't deign to do, to scientists who are indispensable to American science and innovation. The Vilcek Foundation is striving hard to drive home this message.

When the prize program was launched in 2006, the Vilcek Foundation was still operating from my home office. But by then the arrangement was less than satisfactory, given the increasing workload. We needed more people and more space.

We looked into possible options to rent office space, but none appeared attractive. Quarterly royalty payments to the foundation were slowly but steadily rising, reflecting increasing Remicade sales, and by 2006 the foundation had accumulated a small endowment sufficient for the purchase of office space or perhaps even for the acquisition of a small townhouse that could serve as the Vilcek Foundation headquarters. Also, by this time Remicade had been in use for about eight years and the danger that unexpected circumstances might force the FDA to withdraw the license was unlikely. In fact, the FDA was adding new disease conditions to the list of approved indications for Remicade: ankylosing spondylitis in 2004, psoriatic arthritis and ulcerative colitis in 2005, and severe plaque psoriasis in 2006. With these new indications and the prospect of a further increase in Remicade sales, the foundation seemed on its way to financial security.

Our executive director, Rick Kinsel, and Marica started scouting available townhouses and were drawn to a two-story, landmark-protected former carriage house on East Seventy-Third Street between Lexington and Third Avenue, built in 1902. In addition to an attractive Beaux-Arts facade, we liked the building's generous proportions and high ceilings. The downside of acquiring a hundred-plus-year-old building was that it required an extensive renovation. But having to demolish the interior of the building meant that we had the opportunity to fashion it to our needs and our taste. The architect Peter Tow, who had been battle-tested on smaller renovation projects, was our natural choice for the job. By November 2007 we were ready to move into the Vilcek Foundation's new headquarters.

From having no space at all we suddenly had at our disposal around seven thousand square feet of usable space. Within weeks we hired three new employees, two young women to handle the programs and one administrative assistant. It was a completely new experience and we were thrilled.

About half of the space on the main floor of the foundation headquarters is an open space suited for small exhibitions. The gallery space features three arched glass windows facing a handsome rear garden that belongs to our neighbors. Rick came up with plans to use the space for exhibitions of foreign-born artists living in the US. Similar to the prize program, we intended the exhibitions to help immigrant artists become better established and to alert the public to the important contributions made by living foreign-born artists.

Our very first exhibition featured Korean-born artist Il Lee, who uses ballpoint pens to create distinctive minimalist abstract images on paper or canvas. Relatively unknown at the time he presented his work at what we started to call "the Vilcek Foundation Gallery," his work has since been exhibited in many places including the Metropolitan Museum. Paired with Il Lee's work in the first exhibition at our gallery was work by Iranian-born Pouran Jinchi, whose paintings are inspired by Iranian calligraphy.

A total of ten exhibitions have been featured at the Vilcek Foundation Gallery since the first show in 2008. The most popular exhibition, in terms of visits and audience interest, was *LOST*, featuring the contributions of some twenty immigrant and first-generation artists to this wildly popular television show. We also fondly remember a 2009 exhibition by the then relatively unknown Japanese-born artist Ryo Toyonaga, who used clay to create sculptures whose shapes were "driven by an energy welling up from the dark field of his subconscious." Ryo's work too has become much better known since our exhibition; a large retrospective exhibition of his work was held in Eugene, Oregon, in the fall of 2014.

The East Seventy-Third Street building served us well for several years. After about five years we realized that we had outgrown the space and that we needed more room for exhibitions and offices. Recently, the foundation acquired another townhouse on East Seventieth Street between Fifth and Madison avenues that is about 50 percent roomier than the old one. Upon completion of the renovation, scheduled for mid-2016, the foundation will move into the new space and sell the old building.

———

We were happy with our prize program and we had the feeling that it was accomplishing the goal of raising public awareness about the contribution of immigrants to the arts and science. Nevertheless, a couple of years into the program we could not avoid the impression that for most of our distinguished prizewinners the $50,000 (now $100,000) award and the recognition that comes with it was probably icing on the cake. The majority of these people had already been decorated multiple times, and the financial award, as welcome as it might be, was not changing their lives in a profound way. Wouldn't it be appropriate to add prizes for a younger generation of foreign-born scientists and artists, for whom receiving the awards would likely be more significant?

We reached the tentative decision to add prizes for young scientists and artists in 2007, with the goal of a first awards cycle in the spring of 2009. But there were logistical challenges. It was relatively easy to identify candidates for the main prizes, because they were already famous. With the younger candidates it was different; they had to come to us, not the other way around. We needed to institute an effective program of nominations or applications.

The system that eventually emerged was akin to the process of college applications. Applicants would submit a set of biographical data, provide copies of their most important scientific publications or artistic creations, and write short essays with information about their past work, immigration experience, and future professional plans. In order for the first prizes to be awarded in the spring of 2009, we had to be ready for the submission of applications a year in advance. Even with these decisions out of the way we still had our work cut out for us.

One unresolved question was what age limit we would set for the applicants. Marica and Rick wanted the maximum age to be low, perhaps thirty. I argued that thirty may be appropriate for artists, but not for biomedical scientists whose training is lengthy and who often do not find an independent position until they are in their late thirties or older. I suggested a cutoff age of forty, and eventually we reached a compromise, setting the age limit at thirty-eight. (Only later did it occur to us that we could set different age limits for scientists and artists, which we have done in some years.)

And what would we call the new prizes? How would we distinguish them from the main Vilcek Prizes? Catharine Stimpson, then dean of NYU's Graduate School of Arts and Science, suggested "Vilcek Prize for Creative Promise," and the name stuck. We also decided to set the monetary award at $25,000, half of the amount we paid to the main prizewinners at the time.

Two rounds of changes have been introduced since the Creative Promise Prizes were launched in 2009. As of 2013 three Creative Promise Prizes are being awarded annually in each of biomedical science and the arts, and the awards were first raised to $35,000 each, then to $50,000 apiece in 2015.

Despite the logistical challenges, the Creative Promise Prizes have turned out to be hugely successful and rewarding. In the very first year we received 38 applications in biomedical science and 55 in filmmaking. For the 2014 prizes, we received 165 applications in biomedical science along with 134 in design; for the 2015 round of prizes we received 162 in biomedical science and ninety-two in fashion. But the numbers don't tell the whole story. The accomplishments of most of the young applicants are remarkable. And all of our winners are truly exceptional! We could easily award many more prizes without lowering the standards.

The arts field selected for recognition in 2011 was literature, which led the foundation into yet another new venture. We were quite stunned to learn how many great American writers were born outside the US, including many who had not learned English until quite late in their lives. Our young winner that year was the then thirty-three-year-old Ethiopian-born writer Dinaw Mengestu, our first (though not last) African American prizewinner. Dinaw has three acclaimed novels to his credit (*The Beautiful Things that Heaven Bears; How to Read the Air; All Our Names*). After winning our prize, in 2012, he received the much-admired MacArthur Foundation "genius award."

The four finalists were Yugoslav-born Téa Obreht (who later won the Orange Prize for Fiction for *The Tiger's Wife*), Russian-born Ilya Kaminsky (*Dancing in Odessa*), Welsh-born Simon Van Booy (*Everything Beautiful Began After*), and the Vietnamese-born novelist and short-story writer Vu Tran (*Dragonfish: A Novel*).

Jury members told us how difficult it had been to select only one winner and four finalists. One jury member, Arthur Klebanoff, a literary agent and founder of RosettaBooks, had a suggestion: "Since you can't give out more prizes, why don't you publish an anthology with contributions by the twenty or so top-rated applicants?" The idea appealed to us. We did some research and found that there was only one previously published anthology devoted to writings by foreign-born authors living in the United States that focused on a much older generation of authors, most of whom were no longer alive. Spreading the word about the contribution of young immigrant authors to American literature was a perfect project for our foundation.

We had no prior experience in publishing, but our young, literarily inclined staff member Joyce Li was enthusiastic about the project and took on the responsibility for securing the permits from the authors and their publishers or agents, assembling the manuscripts, and arranging for their copyediting, along with the many other related chores. Rick Kinsel, design-conscious as always, secured the assistance of Joe Shouldice, a talented young designer, who devised an attractive, eye-catching cover: each volume had a slit cut into its front cover with a real no. 2 pencil, inscribed with the title of the publication, snugly inserted into it. The first edition of *American Odysseys: Writings by New Americans*, published in 2012, was a private printing; we distributed complimentary copies to cultural institutions and libraries. An attractive paperback trade edition of the same book was published by Dalkey Archive Press in 2013. We are proud of our publishing venture.

In film, the Vilcek Foundation has had a long relationship with the Hawaii International Film Festival (HIFF), a collaboration established in 2007 by Rick (whose long-abiding love of Hawaii also influenced our *LOST* exhibit). The festival, held annually in Honolulu and some other locations in Hawaii, is run by a nonprofit organization dedicated to the advancement of cultural exchanges in the Pacific Rim, in existence for over thirty years. HIFF audiences reflect the diverse multicultural face of Hawaii, and much of America. The main reason for our association with HIFF is their enthusiasm to partner with us in the creation of the annual New American Filmmakers program that features a selection of films made by foreign-born filmmakers—directors, producers, editors, and actors. Like other programs we support, it celebrates the vitality and creativity of immigrants—in this particular case their contribution to American cinema.

In 2013 the New American Filmmakers program featured a selection of five films made by foreign-born filmmakers active in the US, whose national origins spanned—in no particular order—New Zealand, the UK, South Korea, France, Cyprus, and Japan. One of the featured movies, *I Learn America*, filmed in New York City, was codirected by French-born director Jean-Michel Dissard and American-born filmmaker and education expert Gitte Peng. The docudrama chronicles how five immigrant students from Asia, Europe, and Latin America, ranging in age from fifteen to nineteen, struggle to adapt to life in New York City and America. The film movingly relates the efforts of the students to assimilate with the help of strongly supportive teachers and school administrators. The students attend Lafayette High School, a public school in Brooklyn that admits exclusively immigrant children. *I Learn America* is only one of a large number of unconventional, highly deserving films created by immigrant filmmakers participating in the New American Filmmakers program.

———

The Vilcek Foundation owes a great deal to the support of its board members. When Jennifer Olshin and Bruce Cronstein stepped down after years of dedicated service (foundation bylaws stipulate that board members cannot serve for more than six consecutive years), other outstanding individuals took their place. One is the prominent patent attorney Peter Ludwig, who advised NYU and me during negotiations of the original licensing agreement with Centocor.

Current members of the board are Joan Massagué, Richard Gaddes (who retired a few years ago from the position of director general of the Santa Fe Opera), and Christina Strassfield (director and chief curator at the Guild Hall Museum in East Hampton, New York).

The staff of the Vilcek Foundation, currently consisting of nine employees (in addition to Rick, our longtime executive director), is a microcosm of the multiethnic and multicultural population of New York City. Four employees are of Asian descent, with China, Korea, and Indonesia represented. A fifth staff member is half Asian African (his father, born on the African island of Mauritius, is of Indian descent) and half Ecuadorian. The remaining four staff members and Rick are native-born Americans of European descent. Although we don't quite count as staff, Marica and I round off the diverse coterie. Everyone gets along splendidly.

———

By now the Vilcek Foundation has been in existence for fifteen years. Marica and I have asked ourselves many times: "Has it been worth it? Or would it perhaps have been preferable to simply give our money to other charities without assuming the responsibility of establishing and managing a foundation?"

My very superficial account of our foundation's activities in the pre-

Birthplaces of Vilcek Foundation Prizewinners

The numbers on the map correspond to the numbers on the lists of prizewinners shown on the facing page.

Winners of the Vilcek Prizes 2006–15*

Name	Award Year	Field	Country of Birth
1 Joan Massagué	2006	Biomedical Science	Spain
2 Christo and Jeanne-Claude	2006	Fine Art	Bulgaria, Morocco (to French parents)
3 Rudolf Jaenisch	2007	Biomedical Science	Germany
4 Denise Scott Brown	2007	Architecture	Zambia
5 Inder Verma	2008	Biomedical Science	India
6 Osvaldo Golijov	2008	Music	Argentina
7 Huda Zoghbi	2009	Biomedical Science	Lebanon
8 Mike Nichols	2009	Filmmaking	Germany
9 Alexander Varshavsky	2010	Biomedical Science	Russia
10 José Andrés	2010	Culinary Arts	Spain
11 Titia de Lange	2011	Biomedical Science	Netherlands
12 Charles Simic	2011	Literature	Yugoslavia
13 Carlos Bustamante	2012	Biomedical Science	Peru
14 Mikhail Baryshnikov	2012	Dance	Latvia (to Russian parents)
15 Richard Flavell	2013	Biomedical Science	United Kingdom
16 Ruslan Medzhitov	2013	Biomedical Science	Uzbekistan
17 Yo-Yo Ma	2013	Contemporary Music	France (to Chinese parents)
18 Thomas Jessell	2014	Biomedical Science	United Kingdom
19 Neri Oxman	2014	Design	Israel
20 Peter Walter	2015	Biomedical Science	Germany
21 Andrew Bolton	2015	Fashion	United Kingdom

*Two Vilcek Prizes, one in biomedical science and one in the arts, are awarded every year to distinguished foreign-born individuals in recognition of their contributions to US society. The arts field chosen for recognition changes annually.

Winners of the Vilcek Prizes for Creative Promise 2009–15*

Name	Award Year	Field	Country of Birth
22 Howard Chang	2009	Biomedical Science	Taiwan
23 Ham Tran	2009	Filmmaking	Vietnam
24 Harmit Malik	2010	Biomedical Science	India
25 Varin Keokitvon	2010	Culinary Arts	Laos
26 Yibin Kang	2011	Biomedical Science	China
27 Dinaw Mengestu	2011	Literature	Ethiopia
28 Alice Ting	2012	Biomedical Science	Taiwan
29 Michel Kouakou	2012	Dance	Ivory Coast
30 Hashim Al-Hashimi	2013	Biomedical Science	Lebanon
31 Michael Rape	2013	Biomedical Science	Germany
32 Joanna Wysocka	2013	Biomedical Science	Poland
33 James Abrahart	2013	Contemporary Music	United Kingdom
34 Samuel Bazawule	2013	Contemporary Music	Ghana
35 Tigran Hamasyan	2013	Contemporary Music	Armenia
36 Antonio Giraldez	2014	Biomedical Science	Spain
37 Stavros Lomvardas	2014	Biomedical Science	Greece
38 Pardis Sabeti	2014	Biomedical Science	Iran
39 Yasaman Hashemian	2014	Design	Iran
40 Mansour Ourasanah	2014	Design	Togo
41 Quilian Riano	2014	Design	Colombia
42 Sun Hur	2015	Biomedical Science	South Korea
43 Rob Knight	2015	Biomedical Science	New Zealand
44 Franziska Michor	2015	Biomedical Science	Austria
45 Siki Im	2015	Fashion	Germany (to Korean parents)
46 Natallia Pilipenka	2015	Fashion	Belarus (to Ukrainian parents)
47 Tuyen Tran	2015	Fashion	Vietnam

*Vilcek Prizes for Creative Promise are awarded every year to foreign-born artists and biomedical scientists working in the US, who are no more than thirty-eight years of age and who made outstanding contributions at an early stage of their careers. One prize in biomedical science and one in the arts were awarded annually between 2009 and 2012. As of 2013, three Creative Promise prizes each are awarded annually to biomedical scientists and to artists, respectively. The arts field chosen for recognition changes annually.

ceding pages could not include a description of some of the less fulfilling aspects of running a foundation. As is undoubtedly true for managing any organization, whether for profit or not-for-profit, there are responsibilities that are not always fun: tax returns and reports must be filed, financial books must be balanced, investments must be managed, offices must be maintained, and staff members must be hired, supervised, and occasionally laid off. There were times during the fifteen years when the responsibilities that needed to be attended to seemed too much to handle, and moments when inevitably the thought came to Marica's and my minds: "Did we really need this? Is this what we bargained for?"

Before establishing the foundation, Marica and I had separate jobs. On weekdays we parted in the morning and met again for dinner. Of course, we exchanged impressions from our jobs, but our professional responsibilities had not overlapped. The situation has changed since we established our foundation. Suddenly our professional responsibilities *do* overlap. Most of the time, when approaching a problem, we see eye to eye. But inevitably there were times when our views differed. There is of course nothing wrong with that, differences in opinion occur in groups of professional people all the time. Arguments can fly back and forth, and eventually people either agree on a common solution or, if not, then the boss makes a final decision on how things should be handled. Except that for the boss to make an executive decision is not so simple if the dissenting person happens to be his wife . . .

The second thoughts that occasionally drift into Marica's and my heads generally do not persist for long. We feel good about—perhaps—making a small difference in the world.

Chapter Fifteen

Ars Longa, Vita Brevis

In Czechoslovakia the thought of collecting art never entered my mind. I did become a collector of sorts when, as a teenager, with my father's encouragement I became interested in postage stamps, assembling a small collection. The closest I came to collecting art was when I decorated my room in my parents' house with cheap reproductions of the works of Pablo Picasso, cut out from a Polish magazine. Much of Western art of the twentieth century, especially abstract art, was virtually banned in Communist Czechoslovakia as "formalist bourgeois art," so posting images of Picasso's paintings was an expression of my silent opposition to the regime.

Marrying an art historian changed my awareness of the art world. In Czechoslovakia, Marica could not afford to buy art, but already during her studies at the university and later, during her brief career at the print department of the Slovak National Gallery, she assembled a few works given to her by contemporary Czech and Slovak artists that hung on the wall of our living quarters in the house owned by Marica's father.

When we defected from Czechoslovakia we could not, of course, carry with us any of the artwork, but Marica's father managed to remove the pieces from the walls, preventing the authorities from confiscating them when we were sentenced to jail terms and loss of property for the act of illegal emigration. Years later we were able to bring the artwork to New York, and most of those pieces we own to this day.

———

Fast-forward nearly a quarter of a century. Our friend Alvin Friedman-Kien, AIDS researcher, dermatologist, opera lover, and avid eclectic collector of art and antiques of many styles and periods, had been trying for a long time to convince us to join him on his regular summer trips to Santa Fe, New Mexico. His visits there always coincided with the Santa Fe Opera season. For quite a while we resisted Alvin's entreaties. Most of our travels were to places where I attended scientific conferences, often in Europe or Asia; sometimes we would add a few vacation days before or after the meetings. Also, while my parents and Marica's father were alive, we tried to meet them at least once a year, which limited our choice of travel destinations.

When we finally did decide to join Alvin for a four- or five-day visit to Santa Fe in 1988, we were enthralled by the dramatic scenery, exquisite sunsets, unique double rainbows, not to mention the mélange of American Indian, Hispanic, and Anglo cultural influences. Another attraction of Santa Fe was its vibrant art scene and the abundance of interesting galleries and antique shops. And there was the opera featuring five different, usually innovative productions every summer season.

Before returning from our first visit to Santa Fe we promised ourselves that we would come back. And we did, for twenty-two consecutive summers, in Alvin's company.

During our many Santa Fe visits Alvin took great pleasure in introducing us to the local art scene. For many hours during the day he would drag us tirelessly from one gallery and antique shop to the next, squeezing visits in between to an art fair or two and the local flea market.

The visits to galleries, antique shops, art fairs, and flea markets with Alvin were educational. His interests ranged from paintings and sculpture through folk art, American Indian artifacts, African and Asian art, to pre-Columbian art, silver, glass, textiles, furniture, bric-a-brac, and more. In each new gallery or antique shop we would visit, Alvin would quickly scan the surroundings and decide whether there was something of interest to him.

If nothing caught his eye, we would leave. But if he found a piece that interested him, the examination would be intense, with questions to the owner of the shop or gallery and solicitations of our opinions. Often he got carried away. "This is faaah-bu-lous!" he would exclaim. "Look at the perfect surface/shape/color/patina, I have never seen a better example than this one."

Alvin is not only passionate about acquiring things, he loves talking other people into buying objects. "You must get this piece," he would exclaim while we were looking at something in a gallery. Alvin's taste is impeccable, and frequently we did like the objects he thought we would be foolish not to acquire. But during our initial yearly visits to Santa Fe we were on a limited budget, and we could not afford spending three or four thousand dollars on a Navajo blanket, Olmec clay figure, or ancient Indian stone sculpture.

By the time we were returning to Santa Fe for the third or fourth time in the early 1990s our disposable income had grown enough that we were less reluctant to purchase reasonably priced objects we discovered during our forays into galleries with Alvin or on our own. Perhaps inspired by Alvin's style of collecting, initially we did not focus on any particular period, culture, or type of object. Among our early purchases were a small marble bust, probably a Renaissance copy of a Roman emperor's likeness; a pre-Columbian clay pot from the Casas Grandes region of northwestern Mexico; a 1930s drawing by the Santa Fe artist Randall Davey; and a few pieces of contemporary art—paintings, drawings, and sculptures by young New Mexico artists.

Back in New York, we became familiar with the work of Frank Holliday, a contemporary artist who had once worked with Andy Warhol. Rick Kinsel, who at the time worked with Marica at the Metropolitan Museum, purchased a large painting by Frank and then discovered he did not have a big enough wall to hang it on. Marica liked the painting and decided to buy it from Rick as a birthday present for me. Later we befriended Frank, and purchased about half a dozen paintings directly from him. Commercial success has largely eluded Frank, but we continue to admire his work.

At about the same time, Marica and I started acquiring pieces of twentieth-century decorative art and furniture. Initially we bought some pieces of French and Italian art deco furniture from the 1920s and 1930s that gradually replaced our nondescript secondhand furniture acquired from thrift shops during our early days in New York. We became attracted to the art deco esthetic, finding furniture from the first half of the twentieth century easy to live with.

Somewhat later we developed a taste for furniture and decorative objects made in Austria in the very early years of the twentieth century, prior to World War I. This was the period dominated by a group of designers, architects, and artists associated with the *Wiener Werkstätte* (literally "Viennese Workshops"), who created a unique modernist aesthetic. We may have become attracted to these pieces because they reminded us of objects owned by our parents. Most of the pieces we have, including some decorative objects and furniture designed by the Czech-born founder of Wiener Werkstätte, Josef Hoffmann, we acquired during our numerous visits to Vienna.

The Casas Grandes clay pot, purchased during one of our visits to Santa Fe, became the cornerstone of what is now our collection of well over one hundred pieces of pre-Columbian art. Another significant early pre-Columbian acquisition was a stone mask, actually a fragment of a mask, bought for me by Marica as a birthday present. We were originally told the mask dated to the ancient Mexican preclassical Olmec period but, as we later learned, it was very likely of the much more recent Aztec period, which to us did not diminish its beauty or artistic value. Although Marica was enthusiastic about our early purchases, collecting pre-Columbian art eventually became my passion much more than hers.

The scientist in me loved to learn about the many pre-Columbian civilizations and the nuances in the characteristics of their artifacts. And I have gravitated more to stone objects than to ceramics; those that belong to the Mezcala culture, a civilization that existed in the state of Guerrero in southwestern Mexico, particularly appeal to me.

All Mezcala objects that have been preserved are made of stone, most frequently green or gray stone, the great majority of them representing human figures, often in the form of small figurines between two and six inches in height. I am attracted to their simple abstract, geometric features, reminiscent of some twentieth-century cubist creations. The simplest Mezcala figures, presumably the most ancient ones, are little more than stone axes with allusions to a human body, perhaps a neck separating the head from the trunk. The more elaborate ones—still very elementary in design—have legs, arms, hands with a hint of fingers, and a head with simplified geometric images of a nose, ears, and eyes—as if remotely inspired by the facial features of Picasso's *Les Demoiselles d'Avignon*.

The Mezcala people created stone masks, too, as well as images of animals and architectural models that are referred to as "temples." I find the temples remarkable. Ranging in height from about two to perhaps eight inches, they invariably have columns, sometimes stairs leading up to a platform, and a roof. Most of the temples feature only facades, with the naves missing. A fascinating characteristic of some of the temples is the presence of highly stylized human figures, often recumbent—perhaps sleeping, perhaps lifeless corpses—usually placed at the top of the architectural model.

Mezcala art has been compared to the art of another ancient civilization—Cycladic stone objects from the Greek islands of the Aegean Sea dating back to between 3000 and 2000 BCE. The dating of the Mezcala objects is controversial, with estimates ranging from around 1000 BCE to 500 CE. Unlike the great cultures of the Maya or the Aztecs, there are no written records associated with the Mezcala culture.

Another problem with Mezcala art—and, actually, with pre-Columbian art in general—is that there are many fakes. Ceramic objects can be dated using thermoluminescence, but there is no objective method available for dating stone artifacts. Because I have surrounded myself with dozens of pre-Columbian stone objects and because in my professional career I spent hundreds of hours examining images under the microscope, I developed some degree of confidence that by analyzing the stylistic features of a pre-Columbian stone object and by examining its surface with a magnifying glass—paying attention to the age-related abrasions, discolorations, and deposits—I could distinguish fakes from originals. Although I was able to weed out many obvious fakes, with time I learned that there is no completely reliable method to distinguish between genuine stone objects and clever fakes. Some of our objects may not be genuine—a problem also faced by pre-Columbian collections held by famous museums.

Eventually, we switched the focus of our collecting to American modernist art of the twentieth century. Though I still treasure most of the pieces of pre-Columbian art we have assembled, we have not added to the collection for a number of years. Still, it was for many years a fascinating process of learning and researching that ran parallel to my scientific work, enriching my life with greater breadth and perspective.

Santa Fe, which had been the spark and source for our collection of pre-Columbian objects, is also responsible for our collection of American modernist art—by far our most significant foray into the world of art collecting.

During our 2002 summer visit to Santa Fe, Marica and I walked into the Owings-Dewey art gallery, then located on the second floor of a historic building on Santa Fe's main plaza. The gallery, specializing in American art, was a regular destination on our Santa Fe gallery circuit ever since Alvin Friedman-Kien, our longtime friend, introduced us to the owner, Nat Owings, during our very first visit to Santa Fe. While chatting with Nat we noticed an interesting painting hanging on the wall of his gallery office. The painting, a modestly sized landscape with a branchless, leafless forked trunk of a tree at its center, was reminiscent of some early cubist paintings by Picasso and Braque.

We spent some time admiring the painting and discussing it with Nat and his colleague Laura Widmar. I was familiar with Stuart Davis's name, but didn't know a great deal about his work. The painting was expensive by our standards then, and we left without acquiring it. But the image of *Tree* had imprinted itself on our minds and in our hearts. Some time later, we came across the catalog of a 1991–92 Stuart Davis exhibition at the Metropolitan. *Tree* was prominently displayed in it.

When we returned to Santa Fe the next summer, one of our first visits was to the Owings-Dewey gallery. We did not go there with the specific intent of seeing *Tree*. But when we found it hanging on the same wall, in the same spot, we did not let it linger there another year. Soon we acquired other works by Stuart Davis, oil paintings and works on paper. Living with these works turned a budding interest in American modernism into a passion. Quite suddenly we became collectors of American modernist art focusing on works created in the first half of the twentieth century.

Marsden Hartley is another artist prominently represented in our collection, very different in his style of painting and in his life experiences from Stuart Davis. Hartley created his most iconic works in the years immediately preceding the First World War and during the early war years when he lived in Paris and in Berlin. A set of highly ornamental, large, colorful "German officer" paintings created in 1914–15 in Berlin, is described by Patricia McDonnell in a catalog published by the Frederick R. Weissmann Art Museum in Minneapolis as "a panoply of medallions, plumes, helmets, insignia, emblems, flags and banners . . . on parade in Hartley's radical canvasses."

We were fortunate to acquire in 2005—in New York, not Santa Fe—our first two works by Hartley, large charcoal drawings, *Symbol IV* and *Symbol V*, created by Hartley in 1913–14 in Berlin. The drawings relate to Hartley's "Amerika" paintings inspired by Native American symbols, apparently adapted from Hopi Indian kachina dolls, which Hartley admired during his visits to the Museum für Völkerkunde in Berlin. Many years later we purchased a related oil painting, *Berlin Series No. 1*, created in 1913, and one of the very first paintings completed by Hartley in Berlin after moving there from Paris. This work was also part of Hartley's Amerika series. (It is interesting that—like in the title of Franz Kafka's novel—Hartley used the German spelling for America. By coincidence, Hartley's paintings and Kafka's *Amerika* novel were created at about the same time.)

Another significant Hartley painting in our collection, *Portrait Arrangement No. 2* from 1912–13, was painted by Hartley around the time he befriended Karl von Freyburg, the German officer (and possibly Hartley's lover) whose initials appear in many of his paintings of that period. Although our painting reflects mainly Hartley's earlier mystical interests—with a symbolic Indian Buddhist hand gesture at its center—the index finger of the hand touches a circle that encloses the image of a military officer riding atop a white horse, a prequel to his German officer paintings created a year or two later.

Over the years, Marica and I have reached an understanding: we do not acquire an art object unless both of us agree—or at least one of us doesn't veto the purchase. This understanding has worked well for us. Our tastes are actually quite similar, or perhaps have come to be similar after years of collecting art together. However, our temperaments are very different. I get easily carried away. When I like a piece of art my first impulse is "let's get it." Marica is much more levelheaded. She does not like to rush into a decision. Her first reactions are, "Is it a good fit?" "Is this piece really going to enhance our collection?" And, "Where shall we hang it?"

If I had my way, the number of pieces in our collection would by now be unmanageably large and the collection would be much less coherent. On the other hand, if Marica had her way, we would have missed the opportunity to acquire many important pieces over the years. Our collection is indeed a "joint venture." But other people have contributed, especially Rick Kinsel. When Marica and I could not agree whether it was worthwhile to acquire a piece of art, we would ask Rick to play the role of an arbiter. More often than not, we accepted his recommendation.

There were several other instances when Rick's advice was important. Not long after acquiring Davis's *Tree* we purchased another iconic cubist painting by Davis, the large *Still Life with Dial*, which takes its title from the 1920s magazine *Dial*, prominently depicted in the still life. About two years later Rick discovered that there were two other cubist still lifes by Davis available for purchase, of exactly the same size as *Still Life with Dial*, all painted in the same year. Marica and I were initially reluctant to acquire the two related works, *Still Life, Brown* and *Still Life, Red*. "We already have an excellent example of this series, why do we need more?" Rick was able to convince us that owning three related works would make our collection more significant to art history scholars and, besides, the three works would look lovely hung in the same room. He was right. Rick was also instrumental in persuading us to add to our collection several works by Georgia O'Keeffe, another prominent American modernist artist whose work we had initially not included.

As the number of works in our collection grew, we realized that we needed someone to take on the task of preparing a professional catalog. We were fortunate to be able to secure the help of Emily Schuchardt Navratil, a young art history scholar whose specialty is American modernist art. Emily has turned out to be ideal for the job. Besides being a delightful person to work with, she has an uncanny talent for finding and interpreting archival materials. For the first time we had complete information about the provenance and exhibition history of every piece in our collection.

As Emily was busy cataloging, and as the size and quality of our collection was growing, we could not avoid the thought of what would happen to the collection in the long run. *Ars longa, vita brevis*; art endures, life is brief. We are too keenly aware that our collection will survive us, and that we need to make provisions for its future. One option was to select as the beneficiary one or more museums. Many of the best museums in the country would gladly accept our collection as a gift. The problem, from our point of view, is that most objects donated to museums end up in storerooms, from which they emerge only on special occasions. We would prefer to have the art we have assembled available to be seen on a more permanent basis and to keep the collection together.

It was Rick's suggestion that we consider donating our collection to the Vilcek Foundation. Marica and I liked the idea, mainly because the foundation would strive to keep the collection intact. After some thought we decided to earmark all of our American modernist art collection as a promised gift to the Vilcek Foundation. But we were concerned whether the foundation would have enough resources to take proper care of the artworks. Our current building on East Seventy-Third Street has neither sufficient storage facilities, nor the appropriate exhibition space.

But, once the renovation is done by the end of 2015, the foundation's recently acquired townhouse on East Seventieth Street between Fifth and Madison avenues will have much larger and more appropriate exhibition spaces.

Will the Vilcek Foundation have sufficient financial resources to continue caring for the art collection for an extended period of time? As Benjamin Franklin once observed, nothing can be said to be certain except death and taxes. The Remicade royalties assigned to our foundation have helped to establish an endowment that should sustain the foundation for some years to come, even after the expiration of Remicade patents, when no more royalties will be flowing into the foundation coffers. More recently, Marica and I have been adding to the foundation endowment from our personal resources, and we expect to continue doing so.

But even so, the endowment will not last indefinitely. When the time comes, perhaps in fifteen or twenty years, the board at the helm of the foundation will have to decide the ultimate fate of the collection. It seems almost certain that by keeping the collection together in the foreseeable future, the collection will become better known to the world through exhibitions, publications, and online exposure, and its value will continue to grow. Or that at least is our hope and plan.

The cataloging information prepared by Emily has served as the basis for a printed catalog describing and depicting works from our American modernist art collection. The beautifully laid-out book *Masterpieces of American Modernism: From the Vilcek Collection* was published in 2013 by an independent publishing firm in London, Merrell Publishers, known for its dedication to quality of editing, design, and production.

The lead essay was written by William Agee, professor of art history at Hunter College in New York City and author of numerous catalogs, monographs, and articles on American modernism. In it Agee blends equal amounts of expertise and passion to put the works in our collection in the context of the American modernist art movement. The descriptions of the works, placed alongside each of the ninety-eight pieces, illustrated in the book, were written by Lewis Kachur, another expert in the field with many outstanding publications about twentieth-century art to his credit.

Emily Navratil wrote biographies of the twenty artists whose works were included in our collection and prepared a lovely illustrated timeline memorializing important events in the history of American modernist art, especially as they relate to the artists and key works represented in our collection.

Rick, who was responsible for convincing us to go ahead with the publication and for identifying the publisher, wrote an insightful introductory essay that summarizes the story behind our collection. The quality of the final product has exceeded all of our high expectations.

In preparing the catalog, we realized that eight of the twenty artists represented in our collection—40 percent—were born outside the US. The fact came as a surprise to us, because our acquisition decisions were never based on an artist's birthplace. But the revelation that so many artists whose works we own were born outside the US goes to show the significance of foreign-born artists in twentieth-century American art. A happy coincidence, given that the mission of the Vilcek Foundation is to raise awareness of the contribution of foreign-born artists and scientists to America; now it makes even more sense that the foundation is the designated beneficiary of our collection.

Since the publication of the printed catalog, we have made some additions to the American modernist art collection. We have increased the number of the artists represented to twenty-two, and the collection now includes well over 140 paintings, sculptures, and works on paper.

In 2015, sixty-five works from our collection were included in a traveling exhibition entitled *Masterworks from the Vilcek Foundation Collection: From New York to New Mexico*. Organized by Catherine Whitney, chief curator and curator of American art at the Philbrook Museum of Art in Tulsa, Oklahoma, the exhibition debuted at the Philbrook, after which it traveled to the Phoenix Museum of Art and the Georgia O'Keeffe Museum in Santa Fe. The exhibition provided the public with the first opportunity to view a large segment of our American modernist art collection.

———

Apart from the pre-Columbian and the American modernist art collections, Marica and I have also assembled a much smaller collection of Native American pottery. All of the thirty-five pieces in the collection, most of them dating to the late nineteenth and early twentieth centuries, were acquired during our visits to Santa Fe. They range from large ollas, storage jars, and dough bowls to small vessels. One example in our collection is a circa 1880 Acoma Pueblo polychrome olla that stands sixteen inches tall and brandishes intricate geometric designs along with—most interestingly—stylized mice straddling elements of the ornamental geometric pattern.

Several of the American modernist artists in our collection had similar fascinations with Native American pottery. A 1912 painting, *Indian Pottery* by Marsden Hartley, depicts a large Acoma ceramic storage jar along with a wooden carving from British Columbia and a black clay pueblo jar. A 1910 Max Weber painting we own features a ceramic Southwestern Native American bowl next to a Cochiti clay figure.

Our Native American pottery collection and the pre-Columbian art collection too are promised gifts to the Vilcek Foundation.

———

Collecting art has significantly enriched our lives. Not only has our collection provided much pleasure to Marica and me over the years, we enjoy the scholarly aspect of collecting art of a certain style or period, be it pre-Columbian, American modernist, or Native American. I feel a certain pride when I can distinguish between, say, a Nayarit and a Jalisco piece of pre-Columbian pottery, or recognize a George Copeland Ault painting from a distance of fifty feet. Being able to enjoy the art objects we have and learning to appreciate their unique features is enough of a privilege and reward. We have never thought of our art purchases as a financial investment. To bear witness, we have never sold a single piece of art from our collection, even though some pieces have significantly appreciated in value.

There are parallels between the processes of scientific discovery and art appreciation, as indeed there may be in any area in which a person can develop passionate expertise. When you first approach a scientific problem, you begin by reading publications in the field, listening to lectures, discussing the topic with colleagues, then spending hours contemplating and absorbing what you've learned. In art, too, you look, compare, listen, read, and then you look again, until you develop a degree of connoisseurship. There are similarities as well in the lack of predictability in the processes of scientific research and the organic development of an art collection. When Marica and I acquired *Tree*, our first painting by Stuart Davis, we could never have divined that it would become the cornerstone of a significant collection of American modernist art.

How long will the artistic legacy of our collection endure? Only time will tell.

Postscript

Only in hindsight is any legacy made manifest. From my parents I inherited a willingness to participate in the adventure of life. My journey from Bratislava to New York was fraught with unexpected twists and turns: a childhood in the shadow of a frightful war; life under Communist rule and an illegal emigration from my country of birth; a career in research laden with small triumphs and unanticipated twists. From stumbling onto a student research project, and meeting Alick Isaacs and then Albert Sabin, who helped me find the field of research that would become my lifelong passion; to the unexpected opportunity of participating in the development of a drug that revolutionized the treatment of autoimmune diseases and helped millions of people; my future was never preordained.

Along the way I acquired material wealth that I had never aspired to, catapulting me into a new world of philanthropy and art collecting. I did not plan for this particular future.

The title of this book might have been *Life Is Wonderful and Unpredictable*; but who would read a book with so banal a title?

We lived our life. We engaged with the world as deeply as we could, always with love. And in the maze of opportunities life offers, we found the way we made our own.

———

On February 1, 2013, less than five months before my eightieth birthday, I was seated in the ornate East Room of the White House along with twenty-one other scientists chosen to receive either the National Medal of Science or, like me, the National Medal of Technology and Innovation.

247

"Ladies and Gentlemen, the President of the United States."

Within seconds of the announcement President Obama walked into the room—filled already with some two hundred guests.

In his speech, the president talked about the obstacles many of the National Medal recipients had overcome and mentioned four recipients by name:

> One of the scientists being honored today is Jan Vilcek. Jan was born in Slovakia to Jewish parents who fled the Nazis during World War II. To keep their young son safe, his parents placed him in an orphanage run by Catholic nuns. And later, he and his mother were taken in by some brave farmers in a remote Slovak village and hidden until the war was over. And today, Jan is a pioneer in the study of the immune system and the treatment of inflammatory diseases like arthritis.

I only had one thought as I listened to the president's words. "I wish my parents were alive."

Glossary of Medical and Scientific Terms

AUTOIMMUNE DISEASES

There are around one hundred diseases that are suspected or known to be caused by *autoimmunity*. Some of the best-known autoimmune diseases include rheumatoid arthritis, systemic lupus erythematosus, multiple sclerosis, celiac disease, psoriasis, inflammatory bowel disease (Crohn's disease and ulcerative colitis), and type 1 diabetes.

What actually triggers the autoimmune response is generally not known. Autoimmune disorders may be caused by the action of many different components of the immune system, including antibodies, cytotoxic lymphocytes, or a combination of these and other components. Autoimmune diseases tend to be chronic, and inflammation of the affected tissues and organs is often one of the main causes and symptoms of the disease.

AUTOIMMUNITY

The main function of the normal immune system is to protect the organism from invading infectious agents. A normal immune system also plays a role in protection from many cancers. At times, the immune system can go awry and attack healthy tissues and organs inside the body; such action is referred to as autoimmunity. The diseases resulting from the harmful action of the immune system on tissues and organs within its own body are called *autoimmune diseases*.

CYTOKINES

Cytokines are a group of loosely interrelated natural proteins produced in the body, whose function is to regulate a variety of important physiological and pathophysiological functions. Their most important roles are in the regulation of the immune system and in the defense of the organism against infections and cancer. Among the earliest identified cytokines are the *interferons*. Other natural proteins that are counted among the cytokines are substances called interleukins, chemokines, TGF-beta, and members of the large *TNF* (tumor necrosis factor) family. Cytokines are hormone-like substances because, like hormones, they are secreted from cells and act by binding to specific cell surface "receptors," thus eliciting characteristic biological responses. Unlike hormones that are usually produced only by specialized cells (e.g., insulin is made by beta cells of the pancreas), cytokines can be produced by many different cell types and they are usually made in detectable quantities only upon infection or trauma. An important source of cytokines in the body are white blood cells (leukocytes), including lymphocytes, monocytes, and macrophages. Although cytokines generally serve useful functions in host defenses, their excessive production may harm the organism and lead to disease.

Cytokines regulate many immune functions and some cytokines can promote inflammation; thus autoimmunity is often accompanied by a disregulation of cytokine production. Overproduction of some cytokines may be responsible for eliciting and sustaining the inflammatory response underlying autoimmune diseases. The principal cytokines known to promote inflammation in autoimmunity include TNF, IL-1, IL-6, IL-17, and IL-23. Blocking the action of these cytokines by the administration of appropriate monoclonal antibodies or by other means can be beneficial in some—though definitely not all—autoimmune disorders.

INTERFERON

Interferon (sometimes abbreviated IFN) is the name given to a family of natural proteins produced in the organism, usually in response to infection with viruses, bacteria, or other infectious agents. Interferon was first identified in 1957 by the British scientist Alick Isaacs at the National Institute for Medical Research in London, and Jean Lindenmann, a Swiss scientist who was at the time working in Isaacs's laboratory. They observed that influenza virus particles inactivated by heat, though unable to cause infection, were able to suppress, or *interfere with*, the multiplication of fully infectious influenza virus in tissues isolated from chick embryos. The groundbreaking discovery made by Isaacs and Lindenmann was that particles of heat-inactivated influenza virus elicited the production and release of a cellular protein responsible for inhibiting the multiplication of live influenza virus. This protein they termed *interferon*.

It took scientists decades to develop a better understanding of the nature and significance of interferon. It is now known that interferon encompasses a group of many proteins, all of which play roles in the defense against viruses and other infectious agents, in resistance to cancer, and in the regulation of immune functions. On the basis of structural characteristics, interferons are divided into three families termed type I, II, and III inteferons. The family of type I interferons includes IFN-alpha (previously known as leukocyte interferon) and IFN-beta (earlier called fibroblast interferon). Type II interferon is better known as IFN-gamma.

Some interferons are used therapeutically. IFN-alpha is used to treat chronic active hepatitis B and C, and IFN-beta is used to slow the progression of multiple sclerosis. (In hepatitis B, and, especially, in hepatitis C, IFN-alpha treatment is being replaced with other treatments that are more effective and have fewer side effects.) IFN-alpha is also approved as adjuvant therapy in some forms of malignancies; however, its actual use in cancer patients is now relatively rare. IFN-gamma is an important regulator of immune functions, but it has not found wide therapeutic use. The possible medical use of type III interferons is still being evaluated. Like many other cytokines, under some circumstances—especially when produced in excess—interferons can be harmful rather than beneficial.

MONOCLONAL ANTIBODIES
Antibodies are Y-shaped natural proteins produced in the body by specialized white blood cells called B lymphocytes or B cells. The production of antibodies in the organism is triggered by exposure to foreign proteins or carbohydrates, most frequently upon infection with viruses, bacteria, fungi, or parasites. Antibody formation can also be elicited by

injection of a foreign protein or by administration of a vaccine. Antibodies selectively recognize and bind to unique portions of foreign proteins or carbohydrates called antigens (usually parts of invaders such as bacteria or viruses), thereby helping to render the invaders harmless for the organism. Regular antibodies are "polyclonal" because they are produced by thousands of different cells, and consequently they represent mixtures of antibody molecules with very different properties.

In contrast to polyclonal antibodies produced in the human or animal body, monoclonal antibodies are made artificially outside the body by cells that are derived ("cloned") from a single unique antibody-producing cell. The method used to make monoclonal antibodies, pioneered by Georges Köhler and César Millstein at Cambridge University, is to merge ("fuse") a spleen cell from an animal immunized with a selected antigen (commonly a mouse) with a myeloma cell (an immortal cancerous type of an antibody-producing white blood cell), thus producing a hybridoma cell that can be propagated indefinitely while continuing to produce identical molecules of the desired antibody. Scientists have developed many modifications of this original method, permitting industrial-scale production of large quantities of homogeneous, identical monoclonal antibodies. Monoclonal antibodies have many practical applications, including in medical diagnostic procedures; they are also being used successfully for treatment of diseases, such as cancer, infections, and autoimmune disorders.

Because monoclonal antibodies produced in mouse cells are not suitable for administration to humans, scientists have developed methods to reduce or eliminate the presence of mouse sequences in monoclonal antibodies. One such method is the creation of chimeric antibodies by *recombinant DNA* techniques, in which mouse sequences of the original mouse antibody are replaced with human sequences so that the resulting antibody consists predominantly of human sequences. A related process uses recombinant DNA to create "humanized" antibodies—antibodies that are essentially identical to their human variants. Techniques have also become available to produce fully human monoclonal antibodies.

RECOMBINANT DNA AND RECOMBINANT PROTEINS

Recombinant DNA is DNA created in the laboratory by genetic recombination, a process in which segments of DNA derived from different sources are linked together into one DNA strand. Recombinant DNA technology is widely used in research, for example to map and sequence genes and to determine their function.

A very important application of recombinant DNA technology is in the production of proteins. To accomplish that, a DNA encoding a specific protein is linked to some other DNA sequences that, upon insertion into a living cell (bacterium, yeast, or animal cell), will direct the efficient synthesis of the desired protein. The protein ("recombinant protein") can then be isolated from the producing cells or their environment, purified, and "bottled." Much of the biotechnology industry has been built on advances in the production of proteins by recombinant DNA technology. Some examples of proteins that are produced by recombinant DNA technology include human insulin, human growth hormone, hepatitis B vaccine, human papilloma virus vaccine, and human interferon. Most monoclonal antibodies used in medicine are created and produced by recombinant DNA technology.

TNF INHIBITORS

When produced inside the body in excess, TNF can be toxic and harmful. It is now known that TNF is a potent driver of inflammation and a key factor in the pathogenesis (production and development of disease) of numerous chronic autoimmune inflammatory disorders including Crohn's disease, rheumatoid arthritis, psoriasis, ulcerative colitis, and ankylosing spondylitis. These diseases can often be successfully controlled by treatments that inactivate the excess TNF in the body. Several drugs that antagonize TNF have been approved by the federal Food and Drug Administration and by regulatory agencies in many other parts of the world for the treatment of some autoimmune diseases. The first TNF inhibitor approved for treatment in humans in 1998 was Remicade, also known by its nonproprietary (generic) name infliximab. Subsequently, several other TNF inhibitors were approved for therapeutic use in humans. Most of the approved TNF inhibitors are monoclonal antibodies (Humira, Simponi, and Cimzia), and one (Enbrel) is a monoclonal antibody-like artificial construct generated by recombinant DNA technology.

TUMOR NECROSIS FACTOR (TNF)

TNF was first described in a 1975 study by Lloyd Old and colleagues at the Memorial Sloan Kettering Cancer Center in New York City as a protein produced in animals infected with certain bacteria and injected with a bacterial toxin. They observed that the generation of TNF is associated with the regression and death of some malignant tumors in experimental animals, and postulated that TNF plays a role in the body's defense against malignant tumors. Later, when the TNF protein was actually isolated and more fully characterized, it was shown that TNF is responsible for many more activities. It is now known that the principal useful function of TNF in the organism is in the defense against some infectious agents, such as the bacterium that is the cause of tuberculosis. Attempts to use TNF as a therapeutic agent in the treatment of malignant tumors failed mainly because when injected into humans TNF produced too many toxic side effects.

Index

Acknowledgments

When—a few days after my eightieth birthday—I sat down to start writing my memoir, I had only a vague notion about the likeness of the finished product. I knew that my scientific work would serve as the backbone, but I decided early on that I did not want to write a popular science book. Balancing science with the story of my personal life turned out to be one of the major challenges. In the process I relied on the advice of many friends and collaborators.

I am grateful to my wife Marica for her role during the entire process, from the moment I started writing the first draft to the completion of the many revisions that followed. She helped me recall details of events we had experienced together and wisely advised me on what was and was not significant enough to be included in the text. Marica made me realize that parts of my original description of the scientific work were too difficult for readers without science training to comprehend. I am much indebted to my friend and agent Arthur Klebanoff for acting as my guardian angel while I was striving to get the manuscript into a shape suitable for presentation to potential publishers, and, especially, for persuading Dan Simon, the publisher of Seven Stories Press, to give me and my memoir a chance. Arthur recommended that I use the editorial advice of Mina Samuels, a published author. One of the important lessons Mina taught me was "Show, don't tell"; that is, rather than giving in to my tendency to summarize and interpret the described events, I should let the readers reach their own conclusions. Mina helped to improve my prose, influenced by decades of cut-and-dried scientific writing, and make it somewhat more appropriate for the belle-lettres genre of a memoir.

Dan Simon was most generous in applying his legendary editing skills. My original text followed a largely chronological outline. Dan recommended that I take the chapters describing the events surrounding the development of Remicade and place them at the beginning of the volume. "The readers have to be given a chance to grow interested in you before they will want to hear about your childhood." I listened. Dan also reinforced the lesson taught me by Mina Samuels when he said: "Once a story is told, I believe in giving the reader a lot of room to come up with their own interpretations. I also believe that good endings, whether of a paragraph, or a chapter, or a whole book, are precious, and that you have to develop an ear for them so that when that moment comes you're able to know it and to move on from there rather than spoil it by adding extraneous details." I am also indebted to other staff members at Seven Stories Press, especially Lauren Hooker for her enthusiastic and intelligent support. Lauren went out of her way in facilitating the editing and production of this volume. The thorough and methodical copy editor Ibrahim Ahmad uncovered mistakes missed by me and others; to my surprise he was even able to identify spelling errors in the Slovak quotes.

My friend Peter Važan helped to make sure that the descriptions of political events that took place in Central Europe around and during the period of the Second World War were historically correct. Peter unearthed numerous documents relating to the persecution of Jews in Slovakia during that period, which proved helpful. Peter also read and fact checked the text chapter by chapter as I entered it in the computer, catching mistakes and inaccuracies in the process. Dr. Martin Blaser took the time to read an early version of the entire manuscript while he was working on his own book. Marty's most valuable advice was that I simplify the text that recounts the development of Remicade. "Good writing crackles," he said, and "there has to be more story." Dr. Claudio Basilico, who also read an early version, commented that there wasn't enough gossip. I listened and did include more gossip—about Claudio. Joyce Li proofread the entire text twice, each time helping to improve the presentation and making many other smart suggestions. Brian Cavanaugh demonstrated his remarkable expertise by preparing and assembling digital images of the photos included in the memoir. Phuong Pham was inventive and helpful with the preparation of supporting materials. Aaron Pope helped by thoroughly proofreading the galley.

Other friends and colleagues who read either the entire manuscript or selected chapters and made useful comments were Charles Simic, Robert (Bobby) Lamberg, Dr. Jan Zavada, Dr. Joan Prives, Anne Faulkner Schoemaker, David Holveck, Dr. Marc Feldmann, Dr. Lara Marks, Erika Goldman, Franklin Berger, Paul Roy, Emily Navratil, and Rick Kinsel. Rick's powers of persuasion were instrumental in my decision to take up the writing of the memoir. Rick threatened that if I didn't write it, the Vilcek Foundation would hire a ghostwriter to do the job.

There is a much larger group of people to whom I am indebted for the accomplishments described in the memoir. My career in science would not have been possible without the contributions of many talented and hard-working students and collaborators. Only some of them could be mentioned by name in the narrative of the memoir. To make up for the omissions, I am including here complete lists of the students, fellows, visiting scientists, technicians, and administrative assistants who spent time in my lab during the first fifty years of my association with NYU School of Medicine. Their contributions were enormous and essential for my professional career.

Graduate students: Mun H. Ng, Toby Rossman, Douglas R. Lowy (medical student), Sandra Barmak, Brian Berman, Lee W. Mozes, Teresa G. Hayes, Howard Frankfort, Lenore Gardner, Frank Volvovitz, Paul Anderson, Kimberley Pearlstein, David Siegel, Amy Klion (medical student), David Ilson, Vito J. Palombella, Rena Feinman, Luiz F. L. Reis, Jian-Xin Lin, Yihong Zhang, Tae-Ho Lee, Gene W. Lee, Jedd D. Wolchok, Igor C. Oliveira, Peter Sciavolino, Lidija Klampfer, Adam R. Goodman, Ilja Vietor, Paul Schwenger, John Gerecitano, David M. Poppers, Deborah Alpert, Zoltán Szatmáry.

Postdoctoral fellows and visiting scientists: Edward Havell, Ernesto Falcoff, Rebecca Falcoff, Masayoshi Kohase, Shudo Yamazaki, Y.K. Yip, Ioan Sulea, Roy H. L. Pang, Donna Stone-Wolff, Carl Urban, Hanna Kelker, Junming Le, Masafumi Tsujimoto, Dan Aderka, Tadatsugu Taniguchi, Hans-Georg Wisniewski, Deborah Shapiro, Ryutaro Kamijo, Anne Altmeyer, Catalin Mindrescu.

Technicians: Fermina Varacalli, Angel Feliciano, Lyn Gradoville, Irene Zerebeckyj, Dorothy Henriksen-DeStefano, Eve Oppenheim, Barbara Barrowclough, Miriam Pollack.

Administrative assistants: Ellena Kappa, Barbara Dolgonos, Michele Cassano-Maniscalco, Cristy Minton, Ilene Toder-Totillo.

My accomplishments in science would almost certainly not have happened if NYU School of Medicine had not taken the bold step of offering me, sight unseen, a faculty position before my actual arrival as a penniless immigrant on American shores in 1965. As recounted in my memoir, while at NYU I had the good fortune to be able to do exciting basic research, teach (and learn from) some exceptionally talented students, unexpectedly have the chance to contribute to the development of a successful therapeutic drug, and have a great time doing it all.

I am indebted to two institutions in the former Czechoslovakia—now Slovakia—that prepared me for my career in science. The first one is the Medical School of Comenius University in Bratislava. Despite the restrictions imposed by the Communist system, I acquired a decent medical education, and already as a medical student had the opportunity to whet my appetite for research in microbiology and immunology. The second institution in Bratislava that played an important role was the Institute of Virology of the Czechoslovak (now Slovak) Academy of Sciences. During my seven years spent at the Institute of Virology I embarked on studies that remained at the center of my scientific interest for the rest of my career. Arriving at the Institute of Virology as a newly minted medical graduate, I was given a good deal of independence—more than I would have received at an established academic institution in a Western country. That independence could have become a detriment to my development, but somehow it all ended well. For that too I am grateful.

There were many individuals who played important roles in my scientific journey; some are mentioned within the pages of my memoir. I thank them all.